Investigating Groundwater Systems on Regional and National Scales

COMMITTEE ON USGS WATER RESOURCES RESEARCH
WATER SCIENCE AND TECHNOLOGY BOARD
COMMISSION ON GEOSCIENCES, ENVIRONMENT, AND RESOURCES
NATIONAL RESEARCH COUNCIL

NATIONAL ACADEMY PRESS
Washington, D.C.

NOTICE: The project that is the subject of this report was approved by the Governing Board of the National Research Council, whose members are drawn from the councils of the National Academy of Sciences, the National Academy of Engineering, and the Institute of Medicine. The members of the committee responsible for the report were chosen for their special competencies and with regard for appropriate balance.

Support for this project was provided by the U.S. Geological Survey under Contract No. 98HQAG2028.

International Standard Book Number 0-309-07182-8

Library of Congress Catalog Card Number 00-110372

Investigating Groundwater Systems on Regional and National Scales is available from the National Academy Press, 2101 Constitution Avenue, N.W., Washington, D.C. 20418, (800) 624-6242 or (202) 334-3313 (in the Washington metropolitan area); internet <http://www.nap.edu>.

Cover photo, Greer Spring in the Missouri Ozarks, courtesy of Randy Orndorff, U.S. Geological Survey.

Copyright 2000 by the National Academy of Sciences. All rights reserved.

Printed in the United States of America

THE NATIONAL ACADEMIES

Advisers to the Nation on Science, Engineering, and Medicine

National Academy of Sciences
National Academy of Engineering
Institute of Medicine
National Research Council

The **National Academy of Sciences** is a private, nonprofit, self-perpetuating society of distinguished scholars engaged in scientific and engineering research, dedicated to the furtherance of science and technology and to their use for the general welfare. Upon the authority of the charter granted to it by the Congress in 1863, the Academy has a mandate that requires it to advise the federal government on scientific and technical matters. Dr. Bruce M. Alberts is president of the National Academy of Sciences.

The **National Academy of Engineering** was established in 1964, under the charter of the National Academy of Sciences, as a parallel organization of outstanding engineers. It is autonomous in its administration and in the selection of its members, sharing with the National Academy of Sciences the responsibility for advising the federal government. The National Academy of Engineering also sponsors engineering programs aimed at meeting national needs, encourages education and research, and recognizes the superior achievement of engineers. Dr. William A. Wulf is president of the National Academy of Engineering.

The **Institute of Medicine** was established in 1970 by the National Academy of Sciences to secure the services of eminent members of appropriate professions in the examination of policy matters pertaining to the health of the public. The Institute acts under the responsibility given to the National Academy of Sciences by its congressional charter to be an adviser to the federal government and, upon its own initiative, to identify issues of medical care, research, and education. Dr. Kenneth I. Shine is president of the Institute of Medicine.

The **National Research Council** was organized by the National Academy of Sciences in 1916 to associate the broad community of science and technology with the Academy's purposes of furthering knowledge and advising the federal government. Functioning in accordance with general policies determined by the Academy, the Council has become the principal operating agency of both the National Academy of Sciences and the National Academy of Engineering in providing services to the government, the public, and the scientific and engineering communities. The Council is administered jointly by both Academies and the Institute of Medicine. Dr. Bruce M. Alberts and Dr. William A. Wulf are chairman and vice chairman, respectively, of the National Research Council.

COMMITTEE ON U.S. GEOLOGICAL SURVEY WATER RESOURCES RESEARCH

KENNETH R. BRADBURY, *Chair,* Wisconsin Geological and Natural History Survey, Madison
VICTOR R. BAKER, University of Arizona, Tucson
ANA P. BARROS, Harvard University, Cambridge, Massachusetts
MICHAEL E. CAMPANA, University of New Mexico, Albuquerque
BENEDYKT DZIEGIELEWSKI, Southern Illinois University at Carbondale
KIMBERLY A. GRAY, Northwestern University, Evanston, Illinois (through December 1999)
C. THOMAS HAAN, Oklahoma State University, Stillwater (through December 1999)
DAVID R. MAIDMENT, The University of Texas, Austin
DAVID H. MOREAU, University of North Carolina, Chapel Hill (through December 1999)
KAREN L. PRESTEGAARD, University of Maryland, College Park
STUART S. SCHWARTZ, Consultant, San Diego, California
DONALD I. SIEGEL, Syracuse University, Syracuse, New York
VERNON L. SNOEYINK, University of Illinois at Urbana-Champaign
MARY W. STOERTZ, Ohio University, Athens
KAY D. THOMPSON, Washington University, St. Louis, Missouri

National Research Council Staff

WILLIAM S. LOGAN, Project Director
ANITA A. HALL, Project Assistant
RHONDA J. BITTERLI, Editor

WATER SCIENCE AND TECHNOLOGY BOARD

HENRY J. VAUX, Jr., *Chair*, Division of Agriculture and Natural Resources, University of California, Oakland
RICHARD G. LUTHY, *Vice Chair*, Stanford University, Stanford, California
RICHELLE M. ALLEN-KING, Washington State University, Pullman
GREGORY B. BAECHER, University of Maryland, College Park
JOHN BRISCOE, The World Bank, Washington, D.C.
EFI FOUFOULA-GEORGIOU, University of Minnesota, Minneapolis
STEVEN P. GLOSS, University of Wyoming, Laramie
WILLIAM A. JURY, University of California, Riverside
GARY S. LOGSDON, Black & Veatch, Cincinnati, Ohio
DIANE M. MCKNIGHT, University of Colorado, Boulder
JOHN W. MORRIS, J.W. Morris Ltd., Arlington, Virginia
PHILIP A. PALMER (Retired), E.I. du Pont de Nemours & Co., Wilmington, Delaware
REBECCA T. PARKIN, The George Washington University, Washington, D.C.
RUTHERFORD H. PLATT, University of Massachusetts, Amherst
JOAN B. ROSE, University of South Florida, St. Petersburg
JERALD L. SCHNOOR, University of Iowa, Iowa City
R. RHODES TRUSSELL, Montgomery Watson, Pasadena, California

Staff

STEPHEN D. PARKER, Director
LAURA J. EHLERS, Senior Staff Officer
CHRIS ELFRING, Senior Staff Officer
JEFFREY W. JACOBS, Senior Staff Officer
MARK C. GIBSON, Staff Officer
WILLIAM S. LOGAN, Staff Officer
M. JEANNE AQUILINO, Administrative Associate
PATRICIA A. JONES, Study/Research Associate
ANITA A. HALL, Administrative Assistant
ELLEN A. DE GUZMAN, Senior Project Assistant
ANIKE L. JOHNSON, Project Assistant

COMMISSION ON GEOSCIENCES, ENVIRONMENT, AND RESOURCES

GEORGE M. HORNBERGER, *Chair*, University of Virginia, Charlottesville
RICHARD A. CONWAY, Union Carbide Corporation (Retired), S. Charleston, West Virginia
LYNN GOLDMAN, Johns Hopkins School of Hygiene and Public Health, Baltimore, Maryland
THOMAS E. GRAEDEL, Yale University, New Haven, Connecticut
THOMAS J. GRAFF, Environmental Defense Fund, Oakland, California
EUGENIA KALNAY, University of Maryland, College Park
DEBRA KNOPMAN, Progressive Policy Institute, Washington, D.C.
BRAD MOONEY, J. Brad Mooney Associates, Ltd., Arlington, Virginia
HUGH C. MORRIS, El Dorado Gold Corporation, Vancouver, British Columbia
H. RONALD PULLIAM, University of Georgia, Athens
MILTON RUSSELL, University of Tennessee (Emeritus), Knoxville
ROBERT J. SERAFIN, National Center for Atmospheric Research, Boulder, Colorado
ANDREW R. SOLOW, Woods Hole Oceanographic Institution, Woods Hole, Massachusetts
E-AN ZEN, University of Maryland, College Park

Staff

ROBERT M. HAMILTON, Executive Director
GREGORY H. SYMMES, Associate Executive Director
JEANETTE SPOON, Administrative and Financial Officer
SANDI FITZPATRICK, Administrative Associate

Preface

This report is a product of the Committee on USGS Water Resources Research, which provides consensus advice on scientific, research, and programmatic issues to the Water Resources Division (WRD) of the U.S. Geological Survey (USGS). The committee is one of the groups that work under the auspices of the Water Science and Technology Board of the National Research Council (NRC). The committee considers a variety of topics that are important scientifically and programmatically to the USGS and the nation, and it issues reports when appropriate.

This report concerns the work of the WRD in science and technology relevant to assessments of groundwater resources on regional and national scales. The USGS has been conducting scientific activity relevant to groundwater resources for over 100 years and, as summarized in Appendix A, today groundwater-related work occurs throughout the WRD.

Groundwater is a basic resource for humans and natural ecosystems and is one of the nation's most important natural resources. Groundwater is pumped from wells to supply drinking water to about 130 million U.S. residents and is used in all 50 states. About 40 percent of the nation's public water supply and much of the water used for irrigation is provided by groundwater.

Despite the importance of groundwater as one of our most precious natural resources, an organized, effective program to provide an ongoing assessment of the nation's groundwater resources does not exist. With encouragement from the U.S. Congress, the USGS is planning for a new program of regional and national scale assessment of U.S. groundwater resources, thus helping bring new order to its various groundwater resources-related activities. The Survey's senior scientists requested advice in regard to the design of such a program. In response, the committee undertook this study in support of developing an improved program relevant to regional and national assessment of groundwater resources.

Specifically, the Statement of Task to the committee was to "provide guidance to the USGS on development of an improved program relevant to regional and national assessment of ground-water resources." The Statement of Task lists six major topics to be addressed:

1. what constitutes the "regional" and "national" assessment of groundwater resources,
2. how regional studies are chosen (e.g., on the basis of geography, issues, or otherwise),
3. what are emerging issues of regional and national concern,
4. what data and information the USGS should collect and supply and how best to display it,
5. what methods development (e.g., models, geophysical methods, etc.) and research the USGS should pursue for improved future activity, and
6. what are program coordination needs so as to engage outside expertise in priority setting for groundwater studies and to assure communication of results so as to be most useful.

The committee laid out plans for the study in mid-1998 and subsequently met four times before completing this report. At meetings, members were briefed by USGS personnel on a variety of programs and activities. The committee learned about WRD's relevant effort in several hydrologic regions, such as Cape Cod, the southwestern and southern United States, and the Albuquerque and High Plains aquifers. Committee members drafted individual contributions and deliberated as a group to achieve consensus on the content of this report.

As the study proceeded and the committee became more cognizant of USGS activities, productive discussions occurred among committee members and personnel from the USGS and other organizations. This interaction was critical to the success of the project. The committee

Preface   xi

heard from more than 20 USGS staff members and representatives from the U.S. Department of Agriculture, U.S. Environmental Protection Agency, and state and local resource management agencies. The list of individuals providing information to the committee is too long to include in this preface, but we are indebted for the many perspectives and for the information provided. We do wish to single out four individuals from the USGS with whom we interacted throughout the project and thank them for the assistance, information, and cooperation they provided: William M. Alley, chief of the Office of Groundwater; Norman G. Grannemann, coordinator, Groundwater Resource Program; Robert M. Hirsch, chief hydrologist; and Gail E. Mallard, senior hydrologist, who serves as the USGS's continuing liaison with our committee.

The committee hopes that this report will help promote the development of, and appreciation for, improved hydrologic data, information, and knowledge as the USGS supports the nation's effort to manage groundwater resources wisely. The USGS should continue to provide strong, effective, and sustained leadership in this area.

This report has been reviewed by individuals chosen for their diverse perspectives and technical expertise, in accordance with procedures approved by the NRC's Report Review Committee. The purpose of this independent review is to provide candid and critical comments that will assist the authors and the NRC in making the published report as sound as possible and to ensure that the report meets institutional standards for objectivity, evidence, and responsiveness to the study charge. The content of the review comments and draft manuscripts remains confidential to protect the integrity of the deliberative process. We thank the following individuals for their participation in the review of this report, and for their many instructive comments: John D. Bredehoeft, The Hydrodynamics Group; Denise Fort, University of New Mexico; Hugh C. Morris, El Dorado Gold Corporation, C. Kent Keller, Washington State University; John M. Sharp, Jr., University of Texas at Austin; Marios A. Sophocleous, Kansas Geological Survey; and H. Maurice Valett, Virginia Tech.

Although the individuals listed above provided many constructive comments and suggestions, responsibility for the final content of this report rests with the authoring committee and the NRC.

Kenneth R. Bradbury
Chairman, Committee on USGS
Water Resources Research

Contents

EXECUTIVE SUMMARY 1

1 GROUNDWATER AND SOCIETY 6
   A Critical Resource, 9
   An Overdeveloped Resource, 17
   The Necessity for Conjunctive Management, 20
   Conclusions, 24

2 APPROACHES TO SYNTHESIS OF GROUND-
   WATER ISSUES AT THE REGIONAL SCALE 25
   USGS Groundwater Programs-Past and
     Present, 26
   New Opportunities and Mandates, 39
   Proposed Framework for Regional-Scale
     Groundwater Studies, 42
   Conclusions, 47

3 INSTITUTIONAL INTEGRATION AND
   COLLABORATION 48
   External Collaboration, 48
   Internal Collaboration, 57
   Conclusions, 64

4	SCIENTIFIC ISSUES	66
	Aquifer Management, 67	
	Natural Groundwater Recharge, 72	
	Groundwater Quality and Movement in Surficial Materials, 76	
	Groundwater-Surface Water Interactions, 78	
	Groundwater in Karst and Fractured Aquifers, 83	
	Characterization of Subsurface Heterogeneity, 88	
	Numerical Modeling, 91	
	Facilitating Use of Goundwater Information in Decision-Making, 94	
	Conclusions, 98	
5	DELIVERY AND ACCESSIBILITY OF GROUNDWATER DATA	99
	Users of Groundwater Data, 100	
	Content of Groundwater Data, 101	
	Format of Groundwater Data, 102	
	Conclusions, 107	
6	CONCLUSIONS AND RECOMMENDATIONS	108
	Scientific Assessment of Critical Groundwater Issues, 109	
	Regional and National Overviews, 110	
	Access to Groundwater Information, 112	
	Methods Development, 114	

REFERENCES	116
APPENDIX A U.S. GEOLOGICAL SURVEY PROGRAMS THAT SUPPORT GROUND-WATER RESOURCES STUDIES	133
APPENDIX B BIOGRAPHICAL SKETCHES OF COMMITTEE MEMBERS	138

Executive Summary

The U.S. Geological Survey (USGS) has a long and distinguished history in support of regional groundwater management activities, from the construction of water table maps in the late nineteenth century to the Regional Aquifer-System Analysis (RASA) Program of the 1970s and 1980s. However, water resources management has become increasingly complex over time. Large-scale groundwater development throughout the nation has resulted in many ill effects, including lowering of water tables, salt-water intrusion, subsidence, and lowered baseflow in streams, with corresponding ecological damage. Groundwater, surface water, and aquatic ecosystems are now seen to be closely interrelated and can no longer be managed and regulated independently.

Despite the national importance of groundwater, there is little ongoing assessment of the nation's groundwater resources at regional and national scales. The USGS is, however, planning to reshape its Ground-Water Resources Program (GWRP) to focus its various activities in this field. As noted in the Preface, this study was undertaken to assess and provide guidance on efforts to implement such plans.

The committee has concluded that regional groundwater assessment activity of the nature proposed by the USGS should be pursued and that regional and national groundwater assessments should generally have relevance to groundwater sustainability. The management and policy questions that drive regional assessments of sustainability will, in turn, identify and drive the need for regional scientific investigations in fundamental process-oriented groundwater science. The broad topic of sustainability includes the interaction of management decisions (e.g., pumping rates, conjunctive use of groundwater and surface water), resource dynamics (e.g., climatic change, recharge rates), environmental impacts (e.g., streamflow depletion, water quality degradation), and emerging technologies (e.g., aquifer storage and recovery projects). This demand for process-oriented groundwater science to support policy-relevant assessment and management suggests parallel and synthesized approaches, namely regional groundwater *assessments* (i.e., evaluation of the quantity and quality of available groundwater, recharge, extraction rates, etc.) and regional groundwater *science* (i.e., the study of critical processes of regional significance, systematically approached).

This framework can accommodate regional studies of many different types. For example, a groundwater "region" may be defined as a multi-state but geographically contiguous, hydrogeologically distinct area. An example of this is the High Plains aquifer, which was one of the RASA study areas. Alternatively, it may be defined as a discontinuous but widespread aquifer type characterized by a common process or set of processes. An example of this is a karstic aquifer in a temperate climate. One of the accomplishments of the RASA Program was the reconciliation of the hydrostratigraphy of adjacent states. This facilitated the creation of regional groundwater maps and conceptual models of many of the regional flow systems. It is now necessary to broaden that perspective by integrating *processes* as well as properties across regions, and extrapolating the understanding of processes at key sites to larger areas.

The committee has concluded that the following lines of research should be given the highest priority in the context of national and regional groundwater studies:

- Aquifer management: Optimizing groundwater extraction while limiting undesirable effects such as salt-water intrusion, land subsidence, and harm to ecosystems,
- Aquifer Storage and Recovery (ASR) projects: Use of aquifers for repeated storage and recovery of water of varying quality,
- Groundwater recharge: Quantifying rates, spatial locations, and mechanisms of recharge from local to regional scale,
- Surficial aquifers: Evaluating hydrogeology, water-level changes, and water quality changes,
- Interaction of groundwater with surface water: Processes and

Executive Summary

mechanisms in wetlands, rivers, lakes, and coastal areas,
- Characterization of heterogeneous aquifers at large and small scales: Understanding links between geology and hydrogeology, and developing new characterization methodologies, and
- Flow and transport in karst and fractured aquifers.

As part of this research the USGS should continue to explore new modeling techniques that may provide insights into the priority research areas listed above.

Financial resources for regional work should be increasing substantially, but realistically may be limited. The Federal–State Cooperative Water (Coop) Program, which generally focuses on fairly local issues, and the National Water-Quality Assessment (NAWQA) Program together account for nearly 70 percent of the Water Resources Division (WRD) expenditures; the GWRP pilot projects account for only about 2 percent. Because large budget increases are unlikely, a meaningful regional groundwater program must utilize the resources of the NAWQA, Coop, Toxic Substances Hydrology, and National Research Programs. The current Middle Rio Grande Basin and Southwest Groundwater projects have used this management approach with some success. Coordination should include the identification of common data collection, QA/QC, sampling, and archiving protocols to maximize each issue-driven study's contribution to the regional and national groundwater information base.

Of existing programs, the Coop Program is one of the most compatible with regional groundwater studies. Activities of the Coop Program should be aligned with WRD objectives where possible. Because research agendas are driven locally but are competitive, district chiefs can steer activities by selecting the projects most in line with regional objectives.

There are also readily transferable structures between national synthesis in the NAWQA Program (in agreement with NRC, 1994) and the challenges and opportunities for regional and national synthesis in the GWRP. Just as the NAWQA Program was initiated and refined through pilot studies, the Survey's current regional assessments in the Middle Rio Grande basin represent a similar prototype for the development of consistent protocols for regional and national assessment.

The complexity and the multidisciplinary nature of most regional projects argue for collaboration with other divisions within the USGS

(Biological Resources Division, Geologic Division, and Mapping Division) as well as with other federal agencies, state and local governments, universities, and private industry. For example, geologic information (geologic maps, facies analyses, and hydrostratigraphic models) may assist in scaling up the results of a local groundwater study into areas where "hard" hydrogeologic data are sparse or nonexistent.

The USGS should consider formally implementing a steering process for selecting regional and national-scale groundwater issues for study. This should be done through advisory councils operating at the state and national levels, and the Survey should allocate funding for the travel, per diem, and other expenses required to operate these committees. The Survey has taken a similar approach with the Mapping Advisory Council, which aids decision-making for the USGS Mapping Division.

Liaison committees should be established for each study unit, following the model of the NAWQA Program. The committee members should serve as advisors and should consist of state, local, and federal agency water managers, planners, scientists and engineers; university researchers; and representatives from environmental organizations, citizens groups, and other stakeholders. USGS scientists at the district level should be encouraged and rewarded for participating in these and other outreach activities. Summary publications, circulars, and fact sheets appropriate to both technical audiences and decision-makers should be encouraged.

The Survey should also strive to assist in the coordination of groundwater research and data collection among outside groups, such as other federal agencies, state agencies, universities, and the private sector. Although the Survey obviously cannot and should not influence the specific research activities of such organizations, it can provide expert guidance and advice on specific scientific issues and can help shape the national scientific agenda with respect to groundwater research.

In addition to these forms of outreach, however, the GWRP should, along with the other groundwater programs within the USGS, emulate the National Streamgauging Program in moving aggressively to post primary and interpretive data on the Internet. Groundwater data are widely applicable but are expensive to acquire, so their preservation in long-lasting and easily accessible formats should be an integral part of all regional studies.

The proposed medium for many of these kinds of data—the National

Executive Summary 5

Aquifer Data Base—should be well integrated with the existing *National Atlas of the United States* (http://www.usgs.gov/atlas). The GIS-based Atlas already contains a wide variety of viewable and downloadable information from different agencies. It can be viewed at a national, state, or local scale, is organized thematically, and can handle many different data types. Well locations can be plotted and linked to water-level data in a way analogous to the existing system for streamflow data. Ongoing and completed project domains (e.g., the 25 RASA study areas) can be mapped and linked to their data and reports. Maps and model simulation results should be available not only in formats suitable for viewing online, but also in georeferenced formats that can be downloaded and imported into GIS packages.

1

Groundwater and Society

Groundwater, water stored in and transmitted through geologic materials under saturated conditions, is a critical national resource and is often the limiting resource for growth and development. The U.S. Geological Survey (USGS) Water Resources Division (WRD) historically has taken the lead among federal agencies in gathering and distributing groundwater information. The WRD has established a Ground-Water Resources Program (GWRP) to "examine and report on critical issues affecting the sustainability of the nation's ground-water resources." Four activities have been given top priority (USGS, 1998):

- Scientific assessments of critical groundwater issues: Key issues identified by the USGS include groundwater depletion, groundwater–surface water interactions, freshwater/saltwater relations, and groundwater processes in complex geologic environments.
- Regional and national overviews: Ongoing status reports on the nation's water resources.
- Improved access to groundwater data: Easy-to-use Internet interfaces and a national groundwater database.
- Research and methods development: New tools for groundwater investigations.

As noted in the Preface, the committee endeavored in this study to provide general guidance to the USGS on the development of such a program relevant to regional and national assessment of groundwater re-

Introduction

sources and to render a consensus opinion on whether the four proposed activities are consistent with national priorities and the mission of the WRD.

A different National Research Council (NRC) committee (Research Opportunities and Priorities for the EPA, or "ROPE" committee) was given a similar but broader charge in 1995; it was asked to identify and prioritize issues to be addressed in research by the U.S. Environmental Protection Agency (EPA). That committee not only reported on a list of issues, but also made more general recommendations (NRC, 1997a). Of note, the committee recommended that problem-driven research at the EPA be balanced with core research, which emphasizes gaining improved understanding of physical, chemical, and biological processes that underlie environmental systems. This advice seems relevant to the USGS National Groundwater Program. The USGS shares an interest with other agencies, including the EPA, in advancing understanding of such processes. The wording of the USGS priority of "scientific assessments of critical groundwater issues" allows latitude in balancing problem-driven research with core research. However, the committee believes that core research should not be done ad hoc but should be approached explicitly and systematically, as a vital component of a national groundwater program.

The USGS has a long tradition of systematically building the base of understanding of geologic and hydrologic properties on a state-by-state basis, through its district offices. Studies of processes have also been undertaken, but less systematically. The present challenge is to begin working in a multiple-district, regional context, achieving a national synthesis. To some extent this was done in the Regional Aquifer-System Analysis (RASA) Program (Sun and Johnson, 1994) in which the hydrostratigraphy of adjacent states was reconciled and interpreted to create regional maps and conceptual process models. It is now necessary to broaden that perspective by integrating *processes* as well as properties across regions, and extrapolating the understanding of processes at key sites to larger areas. The need for national synthesis is driven by the needs of federal policy and decision-makers. This need is likely to increase as environmental decisions achieve a more integrated global scope.

The committee concurs with earlier NRC reports (e.g., NRC, 1997a) that the task of environmental monitoring and investigation of global-scale issues is too great for any one agency. Interagency cooperation is necessary, as is rapid dissemination of research and data. "Providing accessible groundwater data" and "regional and national overviews" are appropriate priorities and are fundamental to information dissemination to cooperators, decision-makers, and other scientists.

Finally, in addition to providing data and regional and national overviews, the USGS should devote resources to the development of research tools and methods. Highly efficient state-of-the-art tools are needed for measuring environmental variables (e.g., groundwater quality, groundwater levels, subsidence, permeability, and fluxes), for modeling systems and their interactions (e.g., surface water–groundwater interactions), and for interpreting or communicating information for wide use, especially in decision-making. Because the USGS's task in environmental monitoring and basic data gathering is enormous, a strong incentive for efficiency exists. We concur, therefore, that "research and methods development" is a high-priority activity.

Having broadly endorsed the stated priorities, the committee discussed and researched the implications of these priorities for the WRD's activities. The Statement of Task (see the Preface) inspired the eight following questions, which guided the committee's discussions:

1. What are the major groundwater problems and core research needs of the nation? (Chapters 1 and 4)

2. How is the term "region" defined? What constitutes a "regional" assessment? (Chapter 2)

3. How should regional issues be identified and prioritized by the WRD? (Chapter 3)

4. How can issue-driven studies be generalized and synthesized at regional and national scales? (Chapter 4)

5. Can cooperation among the various WRD programs, and among the four USGS divisions, help the WRD to undertake its priority activities? (Chapter 3)

6. What kinds of collaborative arrangements with other local, state, federal, and private institutions would assist the WRD in carrying out regional assessments? (Chapter 3)

7. What tools and methods for streamlining core research and problem-centered research hold the most promise for development and use by the WRD? (Chapter 4)

8. What groundwater information do the clients and cooperators of WRD require, and in what format? How can that information be

Introduction 9

made as widely and rapidly available as possible? (Chapter 5)

Advocating future directions for the USGS WRD first requires an argument for devoting resources to the study of groundwater at a regional scale, and then it requires an argument that the proposed directions are appropriate for the USGS. The remainder of this chapter addresses the first argument; Chapter 2 addresses the historic and future USGS role in groundwater investigations.

A CRITICAL RESOURCE

Drinking and Irrigation Water

Water for drinking and irrigation is perhaps society's most limiting natural resource. Groundwater constitutes only 22 percent of all freshwater used in the United States, but it provides 62 percent of the potable water supply. Roughly 50 percent of the U.S. population and 97 percent of the rural population rely on groundwater as their primary source of drinking water (Figure 1.1). About 40 percent of the nation's public water supply comes from groundwater (Alley et al., 1999; Solley et al., 1998).

In Florida, for example, the Biscayne aquifer is the only source of drinking water for more than 3 million people, about one-quarter of the state's population. In the San Antonio, Texas, area, the Edwards aquifer is the sole source of drinking water for over 1 million people. Similarly, in the Middle Rio Grande basin, the Santa Fe Group aquifer system is the sole source of municipal supply for the city of Albuquerque and many of the surrounding communities, serving about 40 percent of New Mexico's population.

The need for drinking water supplies is not expected to lessen. Based on trends from the past 45 years, water use is expected to grow as population increases, despite per capita declines in use attributed to energy cost, efficiency, conservation and reuse, and regulation (Solley et al., 1998). Especially in the water-poor western states, the persistent search for potable water to fuel urban growth has resulted in pressures on water supplies that may not be sustainable. A variety of water-supply problems have been documented and discussed in NRC reports on small- to large-scale systems (NRC, 1997b,c; NRC, 1998).

FIGURE 1.1 Percentage of population using groundwater as drinking water in each of the 50 states, as of 1995. SOURCE: USGS, 1998.

Groundwater is also the mainstay of agriculture—about 64 percent of all groundwater is used for irrigation. Groundwater provides about 37 percent of irrigation and livestock water supplies nationwide, but in states such as Iowa, Illinois, Mississippi, Missouri, and Wisconsin, this figure is over 90 percent (Solley et al., 1998). As discussed above, agricultural and urban areas are increasingly in competition for the same water resource base.

Streamflow and Ecosystems

In the past few decades, the coupling of surface and groundwater systems has become increasingly apparent. Groundwater–surface water interaction is now recognized as the primary control for such processes as wetland function and riparian habitat maintenance and the geochemical and hydrologic fluxes across the recharge and discharge boundaries of shallow aquifer systems.

Groundwater–surface water interactions involve both matter (in-

Introduction

cluding organisms) and energy, and they occur at all spatial and temporal scales (Winter et al., 1998). Interactions can occur between groundwater and streams (Harvey and Bencala, 1993), lakes (Winter, 1981), wetlands (Siegel, 1988), and estuaries, bays, and coastal areas (Correll et al., 1992; Valiela et al., 1992). Figure 1.2 depicts some of the interrelationships between a stream and its adjacent groundwater reservoir.

Changes in groundwater levels may have major impacts on surface water systems. Rivers literally have been rerouted (Reisner, 1993) or completely dried up (Figure 1.3) by pumping. Conversely, large volumes of stream water may infiltrate into riverbanks at high stage. Bank storage is an important flood-wave attenuation mechanism that is used extensively in engineering hydrology and flood-routing calculations; it also sustains riparian vegetation and may improve surface water quality (Whiting and Pomeranets, 1997).

However, detailed field and modeling studies have also revealed the complexity and variability of shallow groundwater flow paths and their interconnection with streams, lakes, and wetlands.

The direction and magnitude of flows between the two systems can vary rapidly in response to even small changes in boundary conditions (Squillace, 1996; Wondzell and Swanson, 1996; Morrice et al., 1997). These changes, in turn, may have a major impact on critical habitats in these environments. Similarly, riparian buffers are increasingly recognized as playing a crucial role in mitigating the flux of nutrients and flood water for large systems such as the Missouri and Mississippi River basins and Chesapeake Bay. The biological and riparian processes controlling nutrient loads and critical habitats for migratory waterfowl and endangered species in these unique environments are dependent on groundwater–surface water interactions over the range of riparian flow regimes.

Water quality changes in one of the reservoirs may also manifest themselves in the other. Water from a contaminated stream may be drawn into an aquifer by groundwater pumpage (Duncan et al., 1991). This contaminated groundwater may eventually discharge back into surface water (Squillace et al., 1993).

A THREATENED RESOURCE

A limited understanding of the nature of groundwater flow and re-

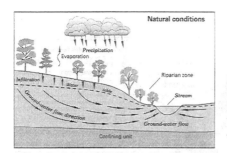

Water is recharged to the ground-water system by percolation of water from precipitation and then flows to the stream through the ground-water system.

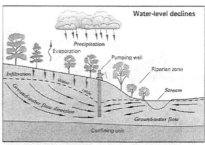

Water pumped from the ground-water system causes the water table to lower and alters the direction of ground-water movement. Some water that flowed to the stream no longer does so and some water may be drawn in from the stream into the ground-water system, thereby reducing the amount of streamflow.

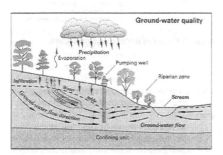

Contaminants introduced at the land surface may infiltrate to the water table and flow towards a point of discharge, either the well or the stream. (Not shown, but also important, is the potential movement of contaminants from the stream into the ground-water system.)

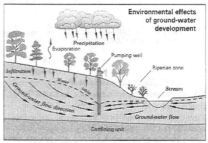

Water-level declines may affect the environment for plants and animals. For example, plants in the riparian zone that grew because of the close proximity of the water table to the land surface may not survive as the depth to water increases. The environment for fish and other aquatic species also may be altered as the stream level drops.

FIGURE 1.2 Effects of groundwater development on a stream–groundwater system. SOURCE: USGS, 1998.

FIGURE 1.3 The Santa Cruz River near Tucson, Arizona, before and after it was dried out by the lowering of The water table by groundwater pumping for the city of Tucson (photographs courtesy of USGS).

has resulted in a legacy of groundwater contamination associated with accidental, improper, or unintended waste disposal. Historically, waste disposal practices relied on landfilling, with little regard for the possible connection between groundwater and the surrounding environment. In some cases, direct injection of liquid waste into aquifers has been utilized as a waste disposal "technology" in residuals management.

By the 1950s, the contamination of the nation's waters by mining, agricultural and industrial chemicals, and sewage had so compromised water supplies that a broad array of federal and state environmental laws and statutes were enacted (NRC, 1993, 1998). These laws have provided substantial protection against further contamination, but many water supplies have been, and continue to be, damaged or threatened by slow-moving contaminant plumes. Also, as noted earlier, groundwater contamination has the potential to reach surface water bodies and the organisms that live in them.

The public recognizes groundwater contamination as a health threat because of highly visible litigation, notably Love Canal, New York (Mazur, 1998), and Woburn, Massachusetts—the site described in the book and movie *A Civil Action* (Harr, 1995). Other books have highlighted the nation's groundwater supply and contamination problems in a manner accessible to nontechnical readers (e.g., Chapelle, 1997; Reisner, 1993). Public concerns over water supply and contamination have led to increased federal and state funding to address fundamental and applied water-related problems. The NRC has prepared a broad array of synthesis and evaluation reports highlighting the nation's evolving understanding of groundwater and surface water contaminant causes, transport, and remediation strategies (e.g., NRC, 1988, 1990, 1993).

Remediation of contaminated industrial sites, often known as "brownfields," is currently a nationwide concern. As our understanding of recharge, contaminant transport, and mixed aqueous phase flow in groundwater grows, the potential scope and impacts of historical waste disposal practices and accidental spills continue to broaden far beyond what was previously thought. For example, slag from the steel industry is ubiquitous in industrial regions and has even been used to reclaim land (Box 1.1). However, water passing through this material may have a pH as high as 12 and may transport trace metals, volatile organic compounds (VOCs), pesticides, and polychlorinated biphenyls (PCBs).

Likewise, nonpoint source pollution, especially from agricultural and urban sources, has become pervasive. For example, groundwater in

Box 1.1
Shallow Aquifer Contamination and Remediation

In the past, many efforts to redevelop contaminated sites were unsuccessful because remediation of the sites to the levels required by regulatory bodies was not economic. For example, an agency might require contaminated groundwater to be cleaned up to meet drinking water standards even though these standards were not met by pristine groundwater, and the aquifer was not even used as a potable water source. This was a common complaint heard by the Illinois Environmental Protection Agency (IEPA) in the early 1990s when the agency evaluated the process by which clean-up objectives were established and the extent to which these objectives posed obstacles to the rehabilitation of abandoned sites. In order to encourage a more cooperative approach to clean-up, the IEPA devised the Tiered Approach to Corrective Action Objectives (TACO), a method for developing remediation objectives for contaminated soil and groundwater that is risk-based and site-specific, protects human health, and considers site conditions and land use (http://www.epa.state.il.us/land/taco/index.html).

TACO provides site owners and regulators with flexible choices for tailoring remediation objectives to local and regional conditions. In the case of groundwater contamination, if it can be shown that the groundwater is not a human exposure route, then remediation objectives and clean-up are no longer needed for the groundwater, and higher amounts of soil and/or groundwater contamination are allowed to remain in place. To be able to exclude a groundwater ingestion pathway, the groundwater in the area of the contamination cannot be consumed as drinking water, and the groundwater must not transport the contamination to a location where it could be consumed. There are a few additional conditions that should be met: free product has to be removed to the extent practical, and any contaminated groundwater discharging to a surface water body must meet current surface water quality standards. A model ordinance was developed for use as an environmental institutional control to prohibit the use of groundwater for potable purposes or the installation and use of new potable water supply wells.

In the city of Chicago such an ordinance was adopted to prohibit the potable use of its shallow aquifer and is relied on to exclude the groundwater ingestion exposure route of contamination. Thus, brownfield remediation and hazardous waste clean-up in Chicago typically ignore the contaminated shallow aquifer. For instance, in 1997 U.S. Steel successfully negotiated the clean-up of its abandoned 567-acre South Works site located on the shore of Lake Michigan in the shadow of the city's downtown. Over the last 100 years, the site was created by filling in the lake with steel mill slag. The groundwater flowing through this material into Lake Michigan, the drinking water source of Chicago, has a very high pH (pH 9–12) and transports a wide range of other contaminants, albeit at low levels. Although the levels of these contaminants are not believed to pose acute health risks to humans or the ecosystem, the long-term fate and the impact of this situation on ecological integrity have not been addressed.

A 1993 USGS study (USGS, 1995) showed that widespread disposal of slag in the Calumet region of Illinois and Indiana has altered groundwater quality primarily by elevating the pH. The direct and indirect consequences of extremely high pH are not well understood. In general, the results of this study indicated that the largest concentration of trace metals, VOCs, semivolatile organic compounds, pesticides, and PCBs were found in samples of the shallow aquifer located in or near industrial areas or areas of waste disposal. In a recent study of the hydrology and water quality of a marsh in the Calumet area by the Illinois State Water Survey (Roadcap, 1999), elevated levels of ammonia-nitrogen, dissolved and suspended solids, and a variety of metals were found to be significantly greater than the maximum values reported for other Illinois natural marshes. This marsh is located adjacent to and downgradient from a cluster of abandoned hazardous waste disposal sites that have contaminated the shallow aquifer. The water chemistry of the system was found to be greatly influenced by the groundwater inflow from the shallow aquifer. Overall, a sound basis for determining the long-term effects of shallow aquifer contamination on the ecological integrity of surface waters is lacking.

an agricultural region covering most of southeastern Washington state has a median nitrate concentration of 9.3 mg/L as nitrogen (the EPA drinking-water standard is 10 mg/L) (Nolan et al., 1998). In the urbanized Coastal Santa Ana basin, volatile organic compounds and pesticides were detected in all monthly and storm samples from surface water-monitoring sites, and in about half of the 20 deep (150-to 300-m) productions wells, some of them in confined settings (Belitz, 1999).

AN OVERDEVELOPED RESOURCE

Large-scale development of groundwater resources has resulted in many undesirable consequences. Three of these—regional subsidence, salt-water intrusion, and resource depletion—are discussed here.

Regional Subsidence

Large-scale development and groundwater extraction can result in irreversible aquifer consolidation and regional subsidence. One of the most infamous examples occurred in the San Joaquin Valley of California, where Poland et al. (1975) estimated that by 1970 subsidence in excess of one foot had affected over 5,200 square miles of irrigable land. Other areas of notable subsidence from groundwater pumping are Houston–Galveston, Texas; Baton Rouge, Louisiana; Santa Clara Valley, California; and the Phoenix area in Arizona (USGS, 1999a). Sinkholes, a particular form of subsidence, are common in the southeastern United States, where pumping from carbonate aquifers has induced collapse.

The cumulative impacts of subsidence can have widespread and unanticipated consequences, including substantial damage to regional infrastructure. Sanitary sewers, for example, are generally designed to flow by gravity to minimize pumping costs. Modest regional subsidence can alter hydraulic grade lines and result in costly damage to water and sewer lines and underground pipelines (NRC, 1995a). Flow in canals can become sluggish or can be reversed entirely. Foundations and roadways can be damaged. Well casings can be crushed by the drag exerted by the subsiding earth. Relatively small cumulative changes in elevation resulting from subsidence alter regional drainage patterns and may significantly change flood risks and drainage in coastal areas and may decrease the flood protection provided by levees and flood-control structures. As an example, Kreitler (1977) estimated that because of subsidence, had Hurricane Carla struck the Houston–Galveston region in 1976 rather than 1961, it would have inundated an additional 25 square miles of land adjacent to Galveston Bay.

Salt-Water Intrusion

Salt-water intrusion in coastal areas is a serious problem, especially along the Atlantic coast (Figure 1.4) where it affects areas from Cape Cod to Miami. In this region, heavy pumping from freshwater aquifers has resulted in the intrusion of salt water, threatening freshwater supplies. Indeed, the aquifers of Brooklyn, New York, were destroyed in the 1930s because of salt-water intrusion induced by excessive pumping, which lowered the water table to 30–50 feet below sea level (Fetter, 1994, p. 367). Incidences of salt-water intrusion into coastal aquifers have been documented in almost all coastal states (USGS, 1998).
Controlling salt-water intrusion is costly and/or management-intensive. For example, water authorities in Tampa, Florida, are planning to build a $95 million desalination plant to replace a portion of their groundwater pumpage and thereby protect their resource from intrusion (Daniels, 2000). In southern California, water managers must continuously maintain a system of hydraulic barriers to intrusion using artificial recharge of storm runoff and reclaimed water combined with pumping wells that continuously remove salt water from the aquifer (http://ca.water.usgs.gov/gwatlas/coastal/la.html). Over 3,000 recharge basins, serving to control drainage and manage groundwater resources, blanket Nassau and Suffolk Counties, Long Island (Ku and Aaronson, 1992). Clearly, salt-water intrusion will continue to be one of the most challenging problems for water managers in coastal regions.

Resource Depletion

The concept of a "safe" or "sustainable" yield of a basin has undergone a long history from the first use of the term "safe yield" by Lee (1915). Although the operational definition may vary from basin to basin (see Chapter 2), sustainable groundwater resource development may generally be viewed as the quantity of groundwater that can be legally extracted from a hydrologic basin over the long term without causing severe economic, social, ecological, and hydrologic consequences. Meaningful investigations of groundwater resource sustainability cannot be limited to county or state boundaries. Accurate quantification of the dynamics of pumping, recharge, consumptive use, and return flows on

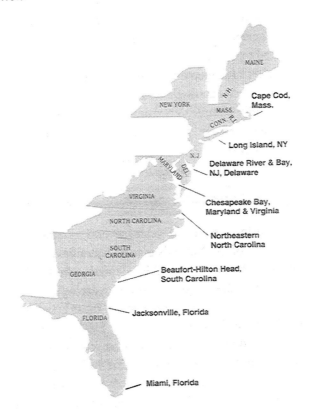

FIGURE 1.4 Areas of salt-water intrusion into freshwater aquifers along the Atlantic coast. SOURCE: USGS, 1998.

regional scales is necessary to evaluate sustainable levels of aquifer development. Characteristic response times and feedbacks between these interrelated processes constrain and characterize both the physical system (e.g., streamflow, depth to water table) and the institutional structures affecting the distribution of benefits and impacts of groundwater use.

Large-scale groundwater depletion has substantial economic costs associated with increases in pumping costs and reduced well yields. However, the economic value of groundwater extraction varies with energy costs and market prices for irrigated agricultural crops. In fact, these economic forces can cause groundwater to be profitably extracted these economic forces can cause groundwater to be profitably extracted beyond the point of sustainability. The cumulative effects of large-scale groundwater development influence, and are influenced by, socioeconomic factors that can effectively transform regional groundwater supplies into a nonrenewable resource.

The High Plains aquifer (Figure 1.5), an important source of water in parts of Colorado, Nebraska, Texas, New Mexico, Kansas, Oklahoma, South Dakota, and Wyoming, is the classic example of this. About 20 percent of irrigated land in the United States is found in this important agricultural region, and about 30 percent of all groundwater used nationwide for irrigation comes from the High Plains aquifer. Between 1940 and 1980, the average water-level decline was about 10 feet, and it exceeded 100 feet in parts of Texas, Oklahoma, and southwestern Kansas (Dugan and Cox, 1994). Pumping lifts and pumping costs have increased in many areas, especially in Texas, making irrigated agriculture less profitable. Since 1980, further declines of over 20 feet over multicounty areas have been common (Gutentag et al., 1984; Zwingle and Richardson, 1993; http://www.ne.cr.usgs.gov/highplains/-hp96_web_-report/hp96_factsheet.htm#-WL8096), although local recoveries have also been noted (Dugan and Sharpe, 1994).

Other aquifers that are being exploited unsustainably ("mined") include aquifer systems of the dry Southwest (e.g., the Albuquerque basin of New Mexico), the Sparta aquifer of Arkansas, Louisiana, and Mississippi, and the Chicago–Milwaukee area aquifer system.

THE NECESSITY FOR CONJUNCTIVE MANAGEMENT

Groundwater depletion, subsidence, salt-water intrusion, and contamination caused by growing demands for municipal, agricultural, industrial, and environmental water may render groundwater a limiting resource for future growth and development. Focusing on these specifics, however, obscures the need to understand—and manage—basins in an integrated manner. The following examples illustrate the integrated approach.

The USGS Middle Rio Grande Basin Study (see Bartolino, 1997b; http://rmmcweb.cr.usgs.gov/public/mrgb/) and Southwestern Ground-Water Resources Project (http://az.water.usgs.gov/swgwrp/Pages/Overview.html) are examples of projects involving fully appropriated surface

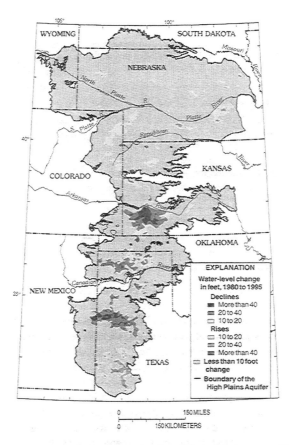

FIGURE 1.5 Water-level declines in the High Plains aquifer, 1980–1995. SOURCE: USGS, 1998.

water systems for which new regional water resources must be developed through conjunctive use of surface water and groundwater. In fact, throughout much of the southwestern United States, surface water is virtually fully appropriated or, in some cases, overappropriated. Significant regional municipal and irrigation demands may directly conflict with riparian environmental requirements, critical to these fragile ecosystems. USGS research has documented the dramatic decline in riparian vegetation associated with groundwater withdrawals (Winter et al., 1998). The variability and robustness of this sensitive riparian environment is also linked to stresses from the hydroclimatic system, for which persistent forcings, such as the regional signature of the El Niño–Southern Oscillation, are recognized as significant sources of interannual variation. The effects of this variability on groundwater recharge and the frequency of extreme events must be taken into account in managing these systems sustainably.

Large-scale water resource development has similarly produced a range of impacts in complex systems like the Florida Everglades. Among other factors, intensive exploitation of groundwater resources to support both agricultural and municipal water demands has resulted in wholesale changes in the regional water balance and has adversely affected the ecology of this unique system. Surface water engineering for flood protection and irrigation demands has imposed anthropogenic variability on the regional hydraulic forcings of the groundwater system. Intricate networks of actively operated canals disrupt the shallow aquifer flow. The delicate coupling of surface and subsurface flow in this low-gradient region critically constrains the restoration opportunities for this system.

As a final example, the Atlantic coastal plain aquifers pose their own distinct challenges for water management. Groundwater–surface water interactions are especially complex in the heterogeneous, unlithified materials that characterize this region, and fluxes into and out of the subsurface are particularly sensitive to changes in land use and surface drainage associated with urbanization. The spatial distribution of recharge and discharge, already highly irregular in these heterogeneous sediments, is made further complex in urban areas by impermeable surfaces, leaky pipes, interbasin transfers, and variable land use. The chemistry of groundwater discharge may also be affected by industrial activity or by intense agricultural uses such as poultry farming. Changes in discharge rates and quality can have major impacts on the health and productivity of riparian, estuarine, and coastal wetlands that provide critical spawning grounds and essential habitat for migratory waterfowl. Finally, long-term pumping around the major cities has led to complex patterns of salt-water intrusion of both deep and shallow aquifers.

Because they are interconnected, groundwater and surface water often behave as one reservoir and should be treated and managed as a single resource (Winter et al., 1998). With regard to water use and allocation, this concept has been recognized for some time and is frequently referred to as "conjunctive use" (Young and Bredehoeft, 1972), and integrated management of the resource is referred to as "conjunctive management." Managers in water-stressed environments recognize the opportunities to enhance the capacity and reliability of regional water supplies through the integrated management of surface water and groundwater. The institutional establishment of recharge districts illustrates the importance of, and opportunity for, integrated management.

The variability, dynamic response, and integrating behavior of groundwater flow systems motivate the need for risk-based planning and evaluation of groundwater resources. This variability has not traditionally been considered in conventional resource evaluation. Resource assessment, recognizing the inherent variability of recharge, flow, and transport processes, is inherently incompatible with the institutional structures that manage water through property rights. Appropriative water law, which treats the rights to water as a static deterministic property right and adheres to the premises of "first in time, first in right" and "if you don't use it, you lose it," raises institutional obstacles to integrated conjunctive management of surface and subsurface water supplies in that it does not adequately account for the inherent spatial and temporal variabilities in groundwater and surface water stocks and flows and for groundwater–surface water interconnections.

Improved understanding of groundwater–surface water interactions enhances the opportunities for joint management and transfers between surface and subsurface supplies through artificial recharge, now commonly synonymous with aquifer storage and recovery (ASR). However, in some Western states, uncertainty in the ability to maintain the right to surface water that is artificially recharged and subsequently extracted as groundwater represents a major institutional obstacle to successful implementation of conjunctive management among competing, and potentially cooperating, water users. For example, until the New Mexico legislature recently amended state water law, anyone artificially recharging surface water would immediately lose the right to that water once it entered the saturated zone and became groundwater. The change was prompted by the city of Albuquerque's desire to implement an ASR program, in which it would recharge excess surface water to replenish groundwater supplies. Prior to the law change, the city would have lost the right to any surface water it recharged to the aquifer.

CONCLUSIONS

Groundwater is critical to the present and future needs of the United States; 130 million people now rely on it for drinking water (USGS, 1998) and have a stake in its sustainability and protection from contamination. But groundwater's role as a component of the hydrologic cycle is equally important. Groundwater has a critical function in maintaining ecosystems, and its connection to surface water dictates that groundwater and surface water must be treated and managed as a single resource (Winter et al., 1998). As society approaches an era that will likely be characterized by great natural and human-induced hydrologic stresses, the USGS is well positioned to maintain its leadership role in monitoring, protecting, and assessing a resource that is essential to the well-being of the nation. The next chapter discusses these roles.

2
Approaches to Synthesis of Groundwater Issues at the Regional Scale

The many groundwater issues summarized in Chapter 1 require an increased emphasis on regional-scale studies. Whereas past regional studies focused mostly on present and future groundwater availability, current issues involve complex interrelated processes that are not adequately characterized by traditional methods of inventorying water resources. How does one conduct a regional study in such a context? Many conceptual and instrumental advances are becoming available to support regional process studies; these are described in Chapter 4. Such advances, however, will merely constitute answers in search of problems without a unified strategy that defines and advances the goals of regional investigations. What is meant by a regional study? Why is it that we need such studies?

This chapter considers alternative approaches to regional studies of groundwater processes. It briefly reviews the groundwater programs and expertise of the U.S. Geological Survey (USGS) and summarizes traditional groundwater resources studies. The customary description of groundwater as a resource is contrasted with a process-based approach to regional groundwater science and regional assessments. Finally, a new approach to regional studies is proposed, based on parallel and synthesized regional groundwater *assessments* (i.e., evaluation of the quantity and quality of available groundwater, recharge, extraction rates, etc.), and regional groundwater *science* (i.e., the study of critical processes of regional significance, systematically approached).

USGS GROUNDWATER PROGRAMS—
PAST AND PRESENT

This section summarizes many of the groundwater programs of the USGS from the early nineteenth century onward, and how these programs have evolved with the changing needs of society.

The Changing Arena of Hydrogeology

The public and scientific arena in which the USGS surface water and groundwater programs function is continually changing. The water resources programs of the Survey initially focused on the investigation of surface water. However, as the development and utilization of groundwater grew following World War I, groundwater investigations became an increasingly important component of its program. In fact, the development of hydrogeology as a scientific discipline is largely the result of work by USGS scientists. In particular, O. E. Meinzer, chief of the Survey's former Division of Groundwater from 1912 to 1946, is generally regarded as the father of modern hydrogeology. USGS scientists have led many of the major scientific advances in the field over the last century (Table 2.1), and these advances have generally paralleled the social and economic needs of the day.

In the early 1900s, bolstered by the growing demand for water and the technological capabilities developed to exploit petroleum resources, regional surveys for discovering, documenting, and developing groundwater became a core element of the Survey's water programs. The widespread availability of economical deep pumps and inexpensive energy made widespread exploitation of ground water resources feasible throughout the Great Plains states and California's Central Valley.

During the first half of the twentieth century, the public needed reliable water supplies to serve a growing and westward-expanding population, with much of this water being consumed for irrigation. These early pressures on groundwater resources revealed many of the issues, including subsidence, groundwater depletion, groundwater contamination, and salt-water intrusion, that continue to define issue-based groundwater investigations conducted by the Survey today. However, most USGS work during this period focused on understanding regional geology and hydrogeology (e.g., Meinzer, 1923) and on an improved theoretical un-

Approaches to Synthesis of Groundwater 27

Table 2.1. Advances in hydrogeology over the last century

Decade	Hydrogeologic advances	Other significant events
1880-1920	-first use of term "hydrogeology" by USGS -understanding of artesian wells Chamberlin (1885) - early groundwater maps (King,1899) -advances in potential theory (e.g. Slichter, 1902)	-World War I
1920-1940	-understanding of regional hydrogeology (Meinzer, 1923) -pumping tests; analytical solutions for groundwater flow (Theis, 1935; Hubbert, 1940)	-Great Depression
1940-1960	-use of electrical resistance analog models -experiments with scaled physical model of groundwater flow -advances in well hydraulics	-World War II
1960-1970	-mainframe computers available -first digital flow models (Pinder and Bredehoeft, 1968) -regional geochemistry (Hem, 1959, Back and Hanshaw, 1965) -groundwater flow systems (Toth, 1962; Freeze and Witherspoon, 1967)	-first satellites program -Vietnam war -Moon landing
1970-1980	-geochemical speciation models (WATEQ) -first digital transport models (Bredehoeft and Pinder, 1973) -groundwater-surface water studies (Winter, 1976) -first practical transport codes (e.g. MOC; Konikow and Bredehoeft, 1978)	-Earth Day; increased public concern for environment -USEPA established; Resource Conservation and Recovery Act (RCRA); Superfund (CERCLA) enacted
1980-1990	-widely-available digital codes (e.g. MODFLOW; McDonald and Harbaugh, 1988) -improved geochemical codes, e.g. PHREEQE -advances in contaminant hydrogeology -USGS RASA studies -practical particle tracking codes, e.g. PATH3D (Zheng, 1992, MODPATH (Pollock, 1989). -improved transport codes, e.g. SUTRA (Voss, 1984); MT3D (Zheng, 1990) -use of natural tracers (tritium, O18)	-personal computers become practical and affordable -contamination events at Love Canal, NY and Woburn, MA -environmental consulting and groundwater remediation become major industries -significant interest in organic contaminants (VOC, NAPL) -internet and world-wide web available
1990-2000	-graphical interfaces for computer codes (e.g. Groundwater Vistas, GMS) -USGS NAWQA studies -increased use of analytic element methods (e.g. Strack, 1989) -inverse model codes (e.g. UCODE, PEST) -innovative groundwater tracers (CFC's) -advances in wetland hydrogeology	-wide use of internet for information transfer -popular book and movie *A Civil Action* describes Woburn events

derstanding of wells and aquifers (e.g., Theis, 1935). The regional groundwater investigations tended to mimic the Survey's mineral resource investigations. The close structural association between the occurrence of mineral deposits and groundwater, combined with the USGS expertise in mapping and geophysical characterization of resource availability and abundance, provided a natural framework for groundwater investigations.

The need for better predictions of basinwide groundwater flow in turn led to the development of mathematical models. The first practical groundwater models were electrical analogs based on networks of resistors and capacitors. As mainframe computers were developed, numerical models came into use, providing the capacity to simulate flow in complex hydrogeologic settings. The USGS developed many of these programs, including MODFLOW (McDonald and Harbaugh, 1988), which may be the most widely used groundwater-flow modeling program in the world today.

At the same time, the inorganic geochemistry of groundwater began to receive more attention. Most of the emphasis was placed on regional geochemistry and the suitability of subsurface water for municipal water use. Limited early investigations of contaminant transport emerged as issue-driven studies of specific incidents associated with industrial spills.

Beginning in the late 1960s, there was increased public awareness of environmental degradation and an associated widespread sense of urgency to protect groundwater from contamination. The public also became aware of issues such as nuclear waste disposal and its related hydrogeologic constraints. These pressures motivated the Survey to go beyond its earlier, relatively isolated district-level evaluations of contaminant transport and to begin conducting fundamental research into the physical, chemical, and biological processes that affect the movement of dissolved constituents through the hydrologic system. In essence, a national demand was created for the technical expertise, data, and ability to synthesize local and regional issues to address national needs. One major outcome was the development in the 1960s, 1970s, and 1980s by USGS scientists of the first practical computer programs for modeling contaminant transport (e.g., Konikow and Bredehoeft, 1978) and geochemical reactions (e.g., WATEQ, Truesdell and Jones, 1974; PHREEQE, Parkhurst et al., 1980).

Approaches to Synthesis of Groundwater 29

Regional Aquifer-System Analysis (RASA) Program

Regional studies earlier in the century described the occurrence of water-bearing formations, their regional extent, and production potential. In the past 30 years, the USGS made more quantitative assessments of the extent, availability, and quality of groundwater resources. The most comprehensive of these was the Regional Aquifer-System Analysis (RASA) Program, which operated from 1978 to 1995. In this program, existing groundwater data were compiled and digital models to simulate regional groundwater flow and response to pumping were constructed for 25 of the most important aquifer systems in the United States. The program's objectives were to define the regional aspects of geology, hydrology, and geochemistry of these systems (Sun and Johnston, 1994), within regions covering thousands to tens of thousands of square miles (Figure 2.1). The completed RASA studies, consisting of a series of reports and maps, constitute the best available synthesis of the regional hydrogeology of the nation. The program was eventually coupled to the National Ground-Water Atlas series begun in 1987 to provide easily accessible regional groundwater resource data to the public. Sun et al. (1997) compiled a bibliography of RASA-related publications.

Although these studies provided an excellent description of general information about the nation's major aquifers, they did not tie that information to many of the issues of sustainability described in Chapter 1. Moreover, although the hydrostratigraphic, hydrogeologic, and geochemical information contained in these studies continues to be useful, demographic changes, population growth, new public attitudes and concerns, and corresponding changes in the regulatory climate drive a need for periodic (5- to 10-year) reassessments of conditions.

National Water-Quality Assessment Program

The National Water-Quality Assessment (NAWQA) Program, implemented in 1991 (after several years of planning) as the RASA Program was nearly completed, is ongoing. Its purpose was to assess water quality distributions and trends, relate these distributions and trends to controlling processes, and evaluate implications for water quality management for a number of large-scale study units (thousands of square

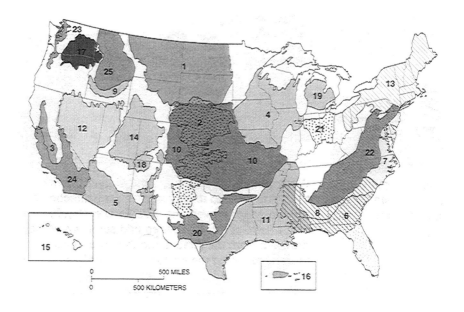

FIGURE 2.1 Regions studied in the RASA Program.

miles) around the nation (NRC, 1994). Typical NAWQA reports describe the environmental settings of large regions, including their geology, soils, climate, land and water use, ecology, and hydrologic systems.

The quality of both surface water and groundwater is inventoried for standard parameters, including dissolved solids, nutrients, trace elements, radionuclides, pesticides, organic chemicals, and bacteria. A key

Approaches to Synthesis of Groundwater 31

component of the program has been the rigorous use of appropriate and consistent sampling and analytical techniques at all sites. These high standards have made NAWQA an especially valuable program for resource assessment.

The NAWQA Program differs significantly from RASA in its focus on water quality and in its inclusion of surface water and biological indicators in addition to groundwater. Because NAWQA is focused on individual study units, a national synthesis of the results is critical, and the National Research Council (NRC) outlined a method to achieve such a synthesis (NRC, 1994). In brief, that synthesis study includes aggregation of data on regional and national levels and aggregation of geographically widespread study units sampled with a consistent methodology; comparisons are then made for spatial and temporal trends and patterns.

Groundwater Monitoring Networks

Understanding the environmental effects of groundwater development on land and surface water resources, and evaluating the overall sustainability of pumping rates, requires long-term monitoring and assessments. The extended time periods are necessary to distinguish transient responses of regional groundwater systems to groundwater withdrawals from other factors such as interannual- to decadal-scale variability in climate and changing land use.

For example, recent increases in groundwater levels in the northern High Plains aquifer appear to be attributable, in part, to an extended period of wet precipitation anomalies from 1980 to 1994 (Dugan and Sharpe, 1994). Although the USGS has been monitoring water levels in the High Plains region for over 60 years, it was only in 1988 that the USGS instituted a formal monitoring program of 8,000 wells to assess annual changes in this important aquifer system (Dugan and Sharpe, 1994). The USGS is the lead agency, since it can monitor or coordinate the monitoring of aquifer systems that underlie two or more states, but the monitoring is performed in cooperation with many other federal, state, and local agencies. A broad network of information is required, because baseline information even from areas that are not presently being heavily exploited may be useful for groundwater resources planning and management.

Other Current USGS Programs

The Federal–State Cooperative Water (Coop) Program, in which state funds are matched by federal dollars (generally on a 50:50 basis) has provided incentive and support for numerous local groundwater resources projects. The cooperative nature of the program ensures stakeholder support on local issues, but the studies have limitations. They tend not to cross state boundaries, and they generally do not address the complex process-related issues that transcend these boundaries. Also, they usually cannot effectively marshal the innovative research appropriate to those process issues.

The USGS Water Resources Division (WRD) also has been active in basic and applied hydrogeologic research through several other programs. Such activities, generally involving the collection of new field data at long-term research sites, have been linked and funded through the WRD's National Research Program, Toxic Substances Hydrology Program, and Water, Energy, and Biogeochemical Budgets (WEBB) Program. The USGS also carries out projects through the Department of Defense (DOD) Environmental Conservation Hydrology program; this work is focused on individual DOD sites. Outside funding also comes from the Department of Energy (DOE), and the Death Valley Regional Ground-Water Flow System Project is one example of a large, regionally integrated study largely supported by DOE funds (see Box 2.1).

Ground-Water Resources Program

Current USGS groundwater assessment is greatly reduced in magnitude from the peak of the RASA program (Figure 2.4). However, a limited number of regional studies are being conducted through the Ground-Water Resources Program (GWRP). As described in Chapter 1, the GWRP was created to examine the sustainability of the nation's groundwater resources through scientific assessments, regional and national overviews, and research and methods development. Its projects are regionally integrated groundwater assessments on a scale smaller than the RASA or NAWQA studies, but larger than most Coop studies. GWRP studies target critical basins or watersheds where important management questions coincide with less-than-adequate technical information.

Box 2.1

The Death Valley regional groundwater flow system (DVRFS) lies in the Great Basin, within a hydrogeologic region characterized by thick sequences of carbonate rocks (see Figure 2.2 and 2.3). These rocks form a generally deep regional aquifer system, which allows inter-basin transfers of ground water from northern and eastern Nevada toward the south and west. The deep regional inter-basin component of the groundwater flow system was recognized by the earliest groundwater investigations conducted by the USGS beginning in 1909. Detailed investigations regarding the role of inter-basin groundwater flow began in the 1960's through a collaborative research effort of the USGS and the State of Nevada. Evidence for inter-basin flow into Death Valley was provided by hydrochemical investigations of spring and well waters in adjacent topographic basins. Soon after this study, hydrogeologists began adopting the concept of deep groundwater flow controlled by large regional fault systems. The advent of underground nuclear testing at the Nevada Test Site increased the national significance of the flow system, and as a result provided for further scientific investigations into its complex hydrogeology.

In the late 1970s and 1980s, the Death Valley region again became the focus of regional groundwater investigations in the Great Basin as part of the RASA program and the DOE-supported Yucca Mountain Project. These integrative studies resulted in both conceptual and numerical models of the region. Each of these studies attempted to model the complex three-dimensional (3D) hydrology and hydrogeologic framework. Inevitably, these initial models involved simplifications of the natural heterogeneity exhibited by the flow system. Truly 3D flow modeling was impractical at the time because the methods for representing the complex hydrogeologic framework were not available. With each new modeling exercise, investigators further developed their understanding of the 3D nature of the flow system.

Most recently, two regional groundwater models of the DVRFS were developed independently by the DOE. One model was developed for the Yucca Mountain Site Characterization Office (DOE/YMSCO) and the Nevada Operations Office Hydrologic Resource Management Program (DOE/NV-HRMP). This was developed by the USGS and documented in USGS WRIR 96-4300. Another was developed for the Nevada Operations Office Underground Testing Area (DOE/NV-UGTA) subproject of the Environmental Restoration (ER) Project by private contractors.

In general, the two models are based upon the same basic hydrologic data set. The geologic data sets differ somewhat in detail of interpretation, but the same general hydrogeologic framework is apparent in both models. Areas of significant disagreement occur in the hydrogeologic framework where data are sparse and results are highly interpretive. Estimates of recharge are also highly interpretive and vary significantly throughout the model domains. These differences appear to affect both groundwater flow paths and flux through the models.

In 1998, the DOE requested that the USGS integrate and enhance these two regional models by developing a "second-generation" regional model, which would evolve as new information and tools are developed. Furthermore, the USGS was encouraged by DOE to cooperate to the fullest extent with other federal, state and local entities in the region, including the National Park Service, the Fish and Wildlife Service, the Bureau of Land Management and Nevada and California counties to take advantage of the benefits of their knowledge and expertise.

As a result, the USGS undertook a six-year scientific program to develop the requested regional groundwater flow model and supporting databases. Short-term objectives include the development of a steady-state model that would simulate regional flow system characteristics and could be used to better constrain local scale flow models of specific underground testing areas or Yucca Mountain. Long-term objectives involve the development of a transient model that could be utilized to evaluate natural or human induced changes to the groundwater flow system. Ultimately, the resulting models and corresponding databases will be available to other federal, state, and local agencies and organizations as a cooperative, groundwater management tool.

The DVRFS project draws upon the technical expertise of scientists throughout the USGS. Specialists in hydrogeologic database design and implementation, three-dimensional geologic modeling and groundwater flow modeling are teamed with experts in the geology and water resources of the southern Great Basin. The project includes over 50 full- or part-time scientists from USGS Geologic Division and Water Resources Division offices throughout the western United States.

The DVRFS study is comprised of six integrated work packages. The *regional hydrogeologic database* work package involves the continuous merging, analysis and evaluation of hydrogeologic spatial data, including reassessment of the hydrologic monitoring network and the gathering of new water levels and spring flow measurements to fill gaps in the database. It also involves documenting, distributing and/or archiving the regional modeling database and making the data available over the Internet.

The *geologic* work package develops a comprehensive geologic model for the region based on existing published geologic data. Products include a regional scale (1:250,000-scale) hydrogeologic map, hydrogeologic sections, and geophysical maps estimating the thickness of Cenozoic deposits and the three-dimensional shape of pre-Cenozoic basement.

The *hydrogeologic framework modeling* work package involves developing a digital 3D representation of the regional hydrogeologic conceptual model, incorporating regional lithologic and structural variations affecting groundwater flow, and evaluating framework model-to-flow model discretization scenarios so that the geology is represented appropriately in the groundwater flow model.

Efforts aimed at reducing conceptual *model uncertainties* are conducted in a work package that provides more precise estimates of hydraulic and hydrologic parameters controlling groundwater flow in the system. These estimates are subjected to a rigorous quality assurance and quality control process that may have been lacking in previous programs. The target parameters include estimated evapotranspiration rates, regional infiltration, historical groundwater withdrawals, and aquifer hydraulic properties. An evaluation of potential groundwater flow paths is also conducted utilizing regional hydrochemical data.

The *groundwater modeling* work package involves the design, construction, calibration, and visualization of the regional flow model. Specific activities are also being conducted to evaluate model results through sensitivity and uncertainty analysis utilizing parameter estimation techniques available in the new USGS groundwater modeling code—MODFLOW2000.

The entire regional groundwater flow system assessment process relies strongly on continued *interaction and collaboration* with the technical representatives of stakeholders in the Death Valley region. These technical representatives from government, academia and the private sector offer alternative approaches and ideas to the USGS scientists. Regular open forums are conducted in local communities throughout the region to provide updates of project progress and to solicit technical input into the hydrogeologic assessment and flow modeling process.

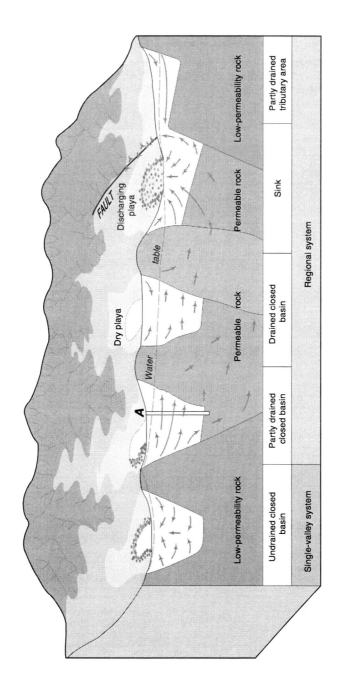

FIGURE 2.2 Generalized ground-water flow patterns occurring in tectonic valleys filled with sediment in the Great Basin. Source: U.S. Geological Survey.

FIGURE 2.3 Spring pool (center) in northern Death Valley. Increased pumping in the basin could eliminate such pools, which provide habitat for the endangered pupfish. View is looking north toward the Montezuma Range. (Photograph courtesy of Frank D'Agnese, U.S. Geological Survey).

One GWRP project that illustrates current thinking regarding regional process-based studies is the Southwestern Groundwater Resources Project. Coordinated from the Arizona district, the project also has study sites in California, Nevada, Utah, and New Mexico, and USGS scientists from many districts and the National Research Program participate. The Southwest is a region of spectacular population growth in which groundwater is especially critical to the economy. In many areas, there is extensive groundwater overdraft, leading to declining groundwater levels, increased recovery costs, decreasing water quality with depth, migration of contaminated water into production areas, subsidence with permanent damage to aquifers, earth fissures, and degradation of riparian habitats. The water budgets associated with overdraft

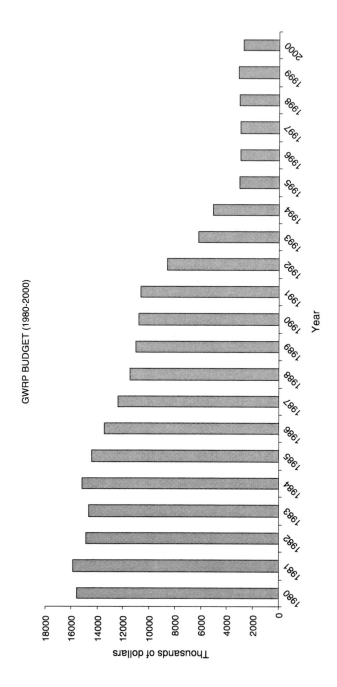

FIGURE 2.4 The annual budget for the GWRP (and its predecessor, RASA) for the last 20 years. Note the decline from more than $15 million in the early 1980s to $3.1 million in 1999. SOURCE OF DATA: U.S. Geological Survey.

are difficult to estimate because of the high spatial and temporal variability of recharge processes and because of climate variability.

The key issues for the southwestern basins are not adequately addressed by water inventories alone. These issues require research into the groundwater processes so that the complexity, variability, and dynamic feedback that govern the systems can be quantified. For example, water use by riparian plants needs to be related to streamflow, which in turn needs to be related to both groundwater conditions and planning and management decisions affecting the sustainability of current and projected resource use. Are the present riparian communities sustainable given anticipated changes in water use? What will be the impact of future changes on endangered species that inhabit the riparian communities? New research approaches will need to be introduced to provide the information on the complex underlying processes that will be necessary to resolve such questions.

The Middle Rio Grande Basin Study (Bartolino, 1997b) is another GWRP project. In this study, the USGS is cooperating with local governments and agencies, including the New Mexico Bureau of Mines and Mineral Resources and the city of Albuquerque, to characterize the hydrogeology of the Albuquerque basin. The project brings together experts in various aspects of hydrogeology and geoscience (geologic mapping, environmental tracers, geophysics, cartography, groundwater modeling, and uncertainty analysis) to understand hydrogeologic problems in the basin. Other GWRP topics include the restoration of the Florida Everglades, salt-water intrusion along the Atlantic coast, and development of a national groundwater database (USGS, 1998).

With the exception of the database, these GWRP projects are being viewed tentatively as prototypes and as a foundation for an expanded program of groundwater studies. An important question addressed in the remainder of this report is whether such efforts represent an appropriate model and scale for future WRD efforts and, if so, how such studies should be prioritized for support.

NEW OPPORTUNITIES AND MANDATES

How might regional needs for groundwater information be met, in view of existing capabilities and potential of the WRD? The breadth of expertise available in the National Mapping, Biological Resources,

Geologic, and Water Resources Divisions of the USGS, along with the Survey's historical role in data collection and management, uniquely positions the USGS to conduct and synthesize local, regional, and national groundwater investigations. Director Charles Groat has pointed out that issues facing the USGS are increasingly complex and interdisciplinary and will tap the capabilities of all four USGS divisions (Molnia, 1999). The Survey's experience with aquifer-based studies (RASA) and place-based, issue-oriented studies (Coop) provides the foundation for the development of process-based scientific and regional assessments of complex problems involving groundwater use for broader purposes than immediate supply.

An Emphasis on Sustainability

New issues are driving a change in priorities in groundwater science and information needs. Whereas regional groundwater assessments traditionally focused on water availability and extraction, new concerns emphasize a complex of issues related to sustainable uses and management. "Sustainability" may have somewhat different meanings, depending on the regional context. For example, a given area may have achieved a long-term, sustainable balance between annual groundwater withdrawals and natural and artificial recharge volumes. However, local areas of excessive withdrawal could well have deleterious effects on riparian habitat. The survival of certain ecosystems may well be controlled by decadal-scale long-term trends, rather than by mean annual conditions. The goal of providing sustainable conditions for future generations may conflict with that of providing optimal supply to current users. Guaranteeing sustainability requires an improved understanding of hydrogeologic processes and the hydrologic budget, and it requires the means of projecting long-term change.

Sustainability implies a basic change in focus from groundwater as an exploitable human resource (the "basin yield" view) to groundwater as a vital part of the complex interrelated processes governing ecosystem health and flow system stability. Such processes include groundwater–surface water interactions, aquifer depletion and associated water-level declines, subsidence and related sinkhole collapse in karstic terrains, and salt-water intrusion. Regions for study will be defined by the effective scale of the dominant process rather than by a physically arbitrary political boundary. Thus, subsidence issues will be associated with a particular sedimentary basin or group of basins unified by the occurrence of subsidence. Salt-water intrusion will be associated with a broad zone of coastal aquifers, and riparian zone impacts will be studied in regions suffering from lowered water tables or near-surface contamination. In short, society has broadened its view of the value of groundwater to include its future use as well as its present use for ecosystem functions. Protecting these values entails a better understanding of hydrogeologic processes in a wide variety of settings at regionally appropriate scales.

Extrapolation to future conditions will require models that appropriately represent measured process rates and dynamics. The limitations of model extrapolations for complex environmental systems will need to be carefully assessed. Advanced measurement methods and advances in modeling need to be balanced with an integrated scientific approach to each study region. Studies will necessarily be interdisciplinary, involving collaborative work with scientists from the other divisions of the USGS. Many of the studies will be of such scope and complexity that expertise from outside the USGS will be required. Collaboration may be facilitated by partnerships with other federal or state science agencies and with universities. This approach to regional investigations requires the integration of regional process-based science and requires regional groundwater assessments oriented to issues of sustainability of current and projected uses, as is discussed later in this chapter.

Alternative Meanings of "Regional"

Groundwater investigations are regional in either of two senses of the word. First, a region may be a hydrogeologically and geographically distinct area encompassing most of one or more states (e.g., the High Plains aquifer). This first type of region has already been defined for the RASA Program. Second, a region may be discontinuous but widespread, encompassing related nonadjacent aquifer systems such as surficial aquifers or coastal aquifers. The discontinuous aquifers of this type of "region" share common processes (e.g., salt-water intrusion, fracture flow, or surface water influence). "Regional" studies in this sense would include some WRD studies traditionally referred to as "topical investigations." The distinction is that many topical investigations have purely local significance. Regional studies of this type may perhaps be viewed as topical investigations that have been optimized for regional applicability. Regional projects of either kind would generally require the cooperation of two or more USGS district offices, and their results would have broad application for water management for a large population.

Problem-based research on regional concerns in the second sense of the word "region" must emphasize processes rather than properties of a specific aquifer. Having characterized the RASA aquifers, the USGS needs to turn its attention to widespread problems having in common not so much geography as process. The USGS has identified the following as the most important problems: groundwater depletion, groundwater-surface water interactions, freshwater/saltwater relations, and groundwater processes in several widespread but complex geologic environments, including karst and fractured rock (USGS, 1998). Information from the RASA studies will help to define the extents of these regions. Chapter 4 examines the appropriateness of these choices.

PROPOSED FRAMEWORK FOR REGIONAL-SCALE GROUNDWATER STUDIES

We propose that future regional-scale groundwater studies have two distinct but complementary components: groundwater assessment and groundwater science.

Regional Groundwater Assessment

An essential policy-relevant form of regional groundwater investigation should focus on *regional groundwater assessments*. In broad terms, these assessments would draw upon the knowledge of the scientific community and the expertise of the Survey to assess the adequacy and sustainability of the quality and quantity of groundwater for its valued uses. Beyond the historically static, descriptive assessment of characteristics such as depth to water table, regional potentiometric maps, and the concentrations of water quality parameters, regional groundwater assessments must integrate process-oriented science to evaluate the sustainability of current and projected trends for hydrogeologically and geographically distinct regional groundwater systems. Of necessity, these regional assessments must be inclusive, incorporating regional

Approaches to Synthesis of Groundwater 43

patterns of water use, ecosystem needs, hydrogeologic and manmade interconnections with surface and subsurface systems, and patterns of growth and development. The demands for regional assessments will challenge the existing tools, information base, and scientific understanding of dominant processes such as recharge, compaction, and groundwater–surface water interactions. Regional groundwater assessment will therefore drive the need for improved regional groundwater science, to refine the understanding of critical processes and feedbacks operating on regional scales.

Regional Groundwater Science

Scientific investigation of processes affecting the integrity of the nation's groundwater may occur in many different geographic regions. These scientific investigations should be undertaken and coordinated as process-based regionalized studies, united by their shared dominant process (e.g., salt-water intrusion, conduit flow). The Survey's unique capability to undertake the study of common processes, contrasted across different hydroclimatic and hydrogeologic settings, provides a robust framework to expand the scientific understanding of the sensitivity and drivers of these processes.

Assessment and science are not independent. Regional assessments quantify and project the impacts from man-made and natural stressors of hydrogeologic systems. These impacts motivate policy and decision-making dynamics and also drive the need for better regional groundwater science. Regional groundwater science builds on the process-based understanding of groundwater systems. Improved process-based groundwater science on regional scales, in turn, supports improved management and policy-making on regional scales. Thus, regional groundwater assessments and regional process-based science form the foundation supporting the synthesis of groundwater investigations on regional and national scales.

Example: Middle Rio Grande

The Middle Rio Grande basin study (http://rockyweb.cr.usgs.gov/public/mrgb/), a GWRP project, is offered as an example of a regional study that may be an appropriate model for future USGS efforts to further our understanding of groundwater at a scale relevant for regional decision-making.

Middle Rio Grande Regional Assessment. The Middle Rio Grande basin study typifies the interdisciplinary scientific and policy issues inseparably woven throughout regional issues related to growth, development, and sustainability of groundwater resource systems. Groundwater withdrawal from the Santa Fe Group aquifer is the primary source of supply for about 40 percent of the population of New Mexico. The sustainability of current and projected patterns of development have far-reaching implications for the regional economy and the ecosystems critically linked through the regional groundwater flow system. More detailed modeling and field investigations have indicated that the available resource, once thought to be an extensive, high-yielding aquifer system, is substantially more limited than had previously been thought (Hawley and Haase, 1992; Thorn et al., 1993).

Moreover, the complexity of the geologic setting and data limitations created considerable uncertainty regarding the interconnections in the aquifer system, the magnitude, extent, and timing of exchange with the riparian system, and the dominant sources of aquifer recharge. Regional groundwater models, calibrated with substantially different assumptions regarding the underlying geologic structure of the aquifer systems, were nevertheless capable of reproducing the observed water levels. However, the parameter estimates associated with these alternate calibrations differed significantly, resulting in dramatic differences in model-computed flow paths and rates of recharge. Critical information needed for regional assessment of the sustainability of current and projected uses thus motivated interdisciplinary research activities to resolve these uncertainties and better understand the fundamental processes influencing the regional groundwater system.

For the Middle Rio Grande–Santa Fe Group aquifer system, the information needed to formulate sustainable policies requires a reliable quantitative description of the regional groundwater flow system, including the hydrogeologic framework as well as patterns and trends in extraction and recharge. The demand for this understanding, in turn, defines critical issues in regional groundwater science that have become the focus of process-oriented groundwater research of regional and national significance.

For example, mountain front recharge in the alluvial aquifers of the Middle Rio Grande basin represents a critical and uncertain control on the regional groundwater system. Research on the mountain front recharge processes is important to this regional study, and leads to the identification of common process-based research on recharge in basin fill aquifers in regions with similar hydrogeologic settings.

Middle Rio Grande Regional Science. The regional assessment of the Middle Rio Grande basin system motivates the need to reliably quantify the recharge, flow paths, surface water interconnections, residence times, and hydrostratigraphic uncertainties in the regional groundwater flow system. One of the critical scientific issues for this regional assessment is accurately characterizing the complex hydrologic connections between the Rio Grande and Santa Fe Group. This need has generated a variety of regional scientific investigations employing gravimetric investigations, digital mapping and geophysical techniques, and the use of environmental tracers to better elucidate the structural controls and flow paths linking the alluvial and aquifer systems.

Similar coordinated investigations have been formulated to clarify the critical processes governing the timing, spatial extent, and rates of mountain front recharge, as well as fluxes between the fluvial and groundwater systems. The indeterminate nature of hydrogeologic parameter estimates derived through the calibration of groundwater flow models has been supplemented by a variety of techniques, including the use of environmental tracers, hydraulic measurement, and thermal pulse tracing, to resolve the streambed interactions. These scientific investigations draw upon and contribute to the scientific expertise and national information base found in the USGS national research program.

Beyond the resulting understanding of aquifer processes, these scientific investigations must also be coordinated to allow their quasi-independent results to be optimally combined, resolving the key process-based questions framed by the regional assessment issues. This need to link scientific investigations to policy-relevant regional assessment issues imposes additional structures, defines critical scales, and requires additional constructs for uncertainty analysis in the scientific research design. In this way, regional assessments further define and shape process-based regional science.

Regionalization

Although the process-based scientific investigations supporting regional assessments will be targeted to the specific hydrogeologic setting, a regional approach to process-based science will expand the applicability of these investigations across hydrogeologic settings. Rather than viewing these local differences as unique settings, a process-based program of regional groundwater science can exploit these differences as cross-sectional intercomparisons, united by their common dominant processes. For example, the detailed investigations of recharge to alluvial aquifers and streambed fluxes in the Middle Rio Grande basin study require scientific understanding of the dominant processes governing the flow between fluvial and groundwater systems. Regional groundwater science focused on common dominant processes, studied across different hydroclimatic and hydrogeologic settings, will expand the scientific understanding of recharge processes common to alluvial aquifers. The USGS's capability to coordinate process-oriented research in different geographic and hydrogeologic settings maximizes the benefits of regional groundwater science.

The domain of regional studies will differ for regional groundwater assessments and regional groundwater science. The scope and extent of regional groundwater assessments will be determined by the spatial domain of the dominant processes and forcings influencing sustainability. The importance of groundwater–surface water interactions will naturally expand the extent of a regional groundwater assessment to include the intersection of the affected watershed and aquifer systems. For example, the growing importance of conjunctive use in regional resource sustainability may naturally extend the limits of a regional assessment to multiple basins, which may be coupled through man-made interconnections establishing interbasin transfers.

The regional domain for groundwater science will be determined by the extent and occurrence of common dominant groundwater processes. For example, karst terrains represent a spatially noncontiguous domain for research on fate and transport of nutrients and microbes in such aquifers. Similarly, investigations of recharge of alluvial aquifers of the Santa Fe Group may best be conducted as part of a regional scientific initiative on recharge processes common to alluvial aquifers of the Lower Colorado (critical to meeting treaty obligations to Mexico), Platte, or Missouri Rivers.

CONCLUSIONS

The USGS is uniquely qualified to conduct and synthesize groundwater investigations on regional and national scales. These investigations will be most effective when structured around process-based rather than resource-based studies. The elements to support these investigations include policy-relevant regional assessments of the sustainability of current and projected patterns of resource use. And finally, the management and policy questions that drive regional groundwater assessments will, in turn, drive the need for regional scientific investigations in fundamental process-oriented groundwater science common to regions with similar dominant processes.

3
Institutional Integration and Collaboration

The scope, complexity, and multidisciplinary nature of many regional water resources problems dictate the need for greater integration of efforts within divisions of the U.S. Geological Survey (USGS) and among stakeholders such as other federal agencies, state and local governments, and the private sector. This chapter discusses the potential for such collaborative efforts.

EXTERNAL COLLABORATION

Collaborative studies with other agencies are not new to the USGS Water Resources Division (WRD) and its predecessors. In fact, in recent years, the WRD has derived a substantial portion of its budget (23 percent in FY 1997) from collaboration with federal agencies.

The WRD has been called on to solve a wide range of groundwater problems, primarily contamination-related, associated with Department of Energy (DOE) and Department of Defense (DOD) activities. WRD projects for the DOE at the Idaho National Engineering and Environmental Laboratory (INEEL; http://water.usgs.gov/pubs/FS/FS-130-97/) and at the proposed high-level radioactive waste repository at Yucca Mountain, Nevada, are examples of such collaboration. The DOD's Environmental Conservation Program (DODEC, formed in 1987) is a cooperative program in which WRD provides technical guidance to the DOD while developing the science of the fate and transport of contaminants in surface and subsurface waters. The WRD has established strong working relationships with other federal agencies as well. Its studies of agricultural pollution in the upper Midwest, conducted through its Toxic Substances Hydrology (Toxics) Program, have been coordinated with both the U.S. Environmental Protection Agency (EPA) and the U.S. Department of Agriculture (USDA).

As a whole, WRD collaborates with a wide range of other organizations (Table 3.1), but the degree to which collaboration occurs varies from district to district. The committee believes opportunities exist that have not been developed in many districts.

Organizations can serve as partners, advisors, or clients in relation to WRD. As partners, collaborators work directly with WRD on projects, collecting data, conducting basic and applied research, assessing water resources for regulatory action, or synthesizing information for consumers at the regional or national level. Collaborators may serve in a more limited capacity as advisors, who steer efforts without actively participating in the project, as clients, who make use of the results of the project, or as both. Primary clients include (1) water managers with operational day-to-day responsibilities for public water supply, protection and distribution, wastewater treatment, reservoir operation, power generation, and flood forecasting, (2) environmental and natural resources managers, regulators, and planners in tribal, local, state, and federal governments and nongovernmental organizations, including burgeoning numbers of watershed groups, and (3) the science and engineering community, including agencies, water supply and treatment industries, mining and energy industries, universities, and engineering and consulting firms. These same organizations and others may be full partners or primarily advisors.

In addition to its formal collaboration with other institutions on specific projects, the USGS must recognize that many other federal, state, local, and private organizations in the United States are engaged in research and data collection on many of the same topics that the USGS groundwater program seeks to address. Many organizations are collecting real-time data on water quality, groundwater contamination, groundwater levels, pumping rates, and a host of other parameters. Some organizations are charged by state law to conduct such studies; others have legal authority to act on water and related issues. The nation's universities and state geological surveys are conducting much fundamental research. One goal of the USGS groundwater program should be to coordinate the efficient gathering of such data and to avoid unnecessary duplication of effort.

TABLE 3.1 Identification and Role of Potential Collaborators with USGS WRD on Groundwater Projects with a Regional or National Component

Potential Collaborator	Type of Collaboration			
	Data Collection	Basic & Applied Research	Regulatory Assessment	Regional or National Synthesis
Other WRD	Partner	Partner	Partner	Partner
Other USGS Divisions	Partner	Partner	Partner	Partner
Other Federal	Advisor, Client	Partner	Partner	Advisor, Client
State	Advisor, Client	Advisor, Client	Partner	Advisor
Local	Advisor, Client	Client	—	—
University/ Water Institute	Partner	Partner	Partner	Partner
Quasi-Gov't.[1]	Partner	Partner	—	—
Private	Client	Client	—	—
Other Stakeholders[2]	Advisor, Client	Advisor, Client	Advisor, Client	Advisor, Client

[1] Includes such entities as water, irrigation, conservancy, water quality, water management, soil and water conservation, and flood-control districts/authorities.

[2] Includes environmental groups, professional organizations, watershed councils, Indian tribes, etc.

Several existing USGS programs with broad participation are described in more detail in this chapter. The Middle Rio Grande basin study (Box 3.1), viewed by many as a prototype for future Ground-Water Resources Program (GWRP) studies, is notable for its Technical Advisory Committee, which helps the many organizations involved in the project to coordinate their programs. The Chesapeake Bay Ecosystem Program (CBEP, Box 3.2) is an example of coordination among all four divisions of the USGS, including the recently incorporated Biological

Box 3.1
Middle Rio Grande Basin Study

The Middle Rio Grande basin study (http://rmmcweb.cr.usgs.gov/public/mrgb/mrgbhome.html) originated in the Federal–State Cooperative Water (Coop) Program, but evolved into a comprehensive 5-year (1995–2000) investigation that stands as a model of collaboration. The basin encompasses a 3,000-square-mile region that includes the Albuquerque, New Mexico, metropolitan area and provides water for 600,000 people, about 40 percent of the state's population. The study's goal is to understand the basin's water resources, in order to support better management decisions by state and local agencies. Much of the effort involves data collection to build a basin model that can be used to test management alternatives. Agencies representing almost every category in Table 3-1 participate in the study as partners, advisors, and clients. They include the Office of the State Engineer, New Mexico Bureau of Mines and Mineral Resources, City of Albuquerque, Bernalillo County, Santa Fe County, City of Santa Fe, EPA, U.S. Bureau of Reclamation, U.S. Army Corps of Engineers, Middle Rio Grande Conservancy District, New Mexico Environment Department, the University of New Mexico, New Mexico Institute of Mining and Technology, Middle Rio Grande Council of Governments, Albuquerque Metropolitan Arroyo Flood Control Authority, and Sandia and Los Alamos National Laboratories (Figure 3.1). Nine Indian nations (the Pueblos of Cochiti, Isleta, Jemez, Laguna, San Felipe, Sandia, Santa Ana, Santo Domingo, and Zia) serve in an advisory capacity, as observers. Other programs within WRD, in particular the National Research Program (NRP), are also actively participating, assisting with numerical modeling and with groundwater tracing and age-dating using environmental tracers and isotopes.

In addition to the WRD, the Geologic and National Mapping Divisions are also actively involved. Activities of the three USGS divisions are coordinated, and to prevent duplication of effort among the involved parties, the Office of the State Engineer has established a Technical Advisory Committee (TAC) for the basin. Membership is drawn from the organizations cited above. Individuals and organizations wishing to perform work in the basin under the auspices of one of the participating agencies must get the approval of the TAC before proceeding. The TAC also has a strong advisory role to the New Mexico state engineer.

FIGURE 3.1 Partners in the Middle Rio Grande basin project meet in formal and informal settings to ensure stakeholder support and participation. The groups represented here include the USGS (WRD and Geologic Division), Office of the State Engineer, Office of the Natural Resource Trustee, and private industry. (Photograph courtesy of James R. Bartolino, USGS.)

The New Mexico WRD district office has hosted a number of formal and informal gatherings of workers and stakeholders, including three workshops featuring papers and discussions. Compendia of extended abstracts and short papers are published as USGS Open-File Reports (see Bartolino, 1997a; Slate, 1998). The USGS also published a fact sheet (Bartolino, 1997b) and maintains a Web site (http://rmmcweb.cr.-usgs.gov/public-/mrgb/mrgbhome.html) for the study.

Box 3.2
USGS Chesapeake Bay Ecosystem Program

The Chesapeake Bay has been negatively impacted over long time scales by nutrient enrichment, influx of sediment and toxic substances, and overharvesting of shellfish and finfish. It now suffers from hypoxia, algal blooms, toxic microbes, and reduced submerged aquatic vegetation (SAV). The USGS is one of several agencies that participate in efforts to protect and preserve the Bay ecosystem.

The USGS contributes to the effort through its Chesapeake Bay Ecosystem Program (CBEP), funded through the USGS Ecosystem Program since 1996. Like other USGS ecosystem programs (e.g., San Francisco Bay, South Florida, and the Mohave Desert), the CBEP is a multiyear, multidisciplinary program. Its objectives are (1) to determine the response of water quality and selected living resources of the bay watershed and estuary to changes in nutrient and sediment inputs and to climatic variability over various scales, (2) to define and evaluate the responses of water quality and living resources to changes in nutrient and sediment sources and climatic variability, and (3) to provide resource managers with the management implications of the scientific findings, especially with respect to nutrient-reduction strategies.

Although the program is coordinated out of the district office in Maryland, CBEP is a "place-based program" with participation from scientists from district offices in Virginia, Pennsylvania, and elsewhere, and from the National Center. Groundwater research forms a significant part of this program, including regional studies on the age, flow paths, residence times, and nitrate concentrations of groundwater (Bachman et al., 1998; Focazio et al., 1998). Other WRD programs are active here as well. The National Water-Quality Assessment (NAWQA) Program, for example, has developed an extensive database of nutrients, organic compounds, and metals in both groundwater and surface water systems. The Chesapeake Bay River Input Monitoring (RIM) Program, a cooperative program, analyzes long-term trends in concentrations of nutrients and suspended material entering the tidal waters.

The other divisions of the USGS participate as well. The National Mapping Division uses digital orthophoto quadrangles to track long-term land-use changes and provides digital line graphs for watershed delineation and surface water modeling. Geologic Division scientists collect sediment cores to infer historical trends in habitats, nutrients, salinity, and sedimentation and to study the paleoenvironmental and geologic history of the bay. The Biological Resources Division is studying carbon and nitrogen cycling across trophic levels, and it is studying the relationships among land use, nutrients, microbes, and fish health.

Interactions with other agencies are also extensive. Land-use maps are made in cooperation with the EPA, National Oceanic and Atmospheric Administration (NOAA), National Aeronautics and Space Administration (NASA), and the University of Maryland at Baltimore County. The analysis of nutrient trends is being performed in partnership with the Susquehanna River Basin Commission, the Maryland Department of Natural Resources (MD DNR), and the Virginia Department of Environmental Quality (VA DEQ). Work on SAV and associated water quality involves the Interstate Commission on the Potomac River Basin, the Bay Program, and the Alliance for the Chesapeake Bay.

The CBEP has its own source of funding from the USGS Ecosystem Program. These funds are leveraged to bring the resources of the various entities to bear on the program. For example, the program funded an additional investigator for the River Input Monitoring Program to complement assessment of water quality changes in rivers feeding the bay (paid for through the Coop Program) with an understanding of the science behind these changes. Similarly, while salaries of scientists investigating the age and nutrient dynamics of shallow groundwater are borne by the National Research Program, most of the costs of the associated field and laboratory work are funded by the CBEP.

The program is coordinated on two levels within the USGS. The Chesapeake Bay Resource Manager Team, composed of program managers, coordinates programmatic issues. The Interdivisional Technical Team, composed of the lead scientists of the various programs, tries to assure that data sets are compatible and mutually supporting. To coordinate with external groups, Survey scientists are on the various subcommittees of the Chesapeake Bay Program, the interstate regional partnership that directs much of the restoration effort in and around the bay.

SOURCES: http://mapping.usgs.gov/mac/chesbay/,
http://wwwdmdtws.er.usgs.gov/chesbay/overview_cbep.html,
http://de.usgs.gov/publications/fs-124-97/,
http://mapping.usgs.gov/mac/chesbay/results.html,
http://dc.usgs.gov/projects/md130.html,
http://geology.er.usgs.gov/eespteam/ches/bayhome.html,
http://www.chesapeakebay.net/ [accessed April 2000], and Scott Phillips, USGS, personal communication., April 2000)

Institutional Integration and Collaboration

Resources Division. The synergy of the CBEP is enhanced by a funding mechanism that leverages and interweaves ongoing programs of the various divisions.

Rationale and Benefits

Collaboration has led to significant contributions to groundwater science and has led to the building of databases for water resource investigations. As water science problems become more inter- and multidisciplinary, WRD collaboration and integration with other organizations will become even more important (NRC, 1991a). Investigations into problems addressed by other agencies engage WRD scientists in new and challenging areas of research and provide settings in which new theories, computer models, and field investigation techniques may be tested. Through these interactions, the WRD would:

- augment and leverage its research funds and programs,
- gain access to research being conducted elsewhere,
- retain its cutting-edge science,
- identify talented individuals (e.g., graduate students) as potential employees,
- stay attuned to current research needs, especially at the local and state levels,
- identify future research needs, especially at the local and state levels, and
- generate good will and political capital.

Conditions for and Obstacles to Collaboration

Collaboration in providing the science and databases to manage regional groundwater resources requires communication between WRD and stakeholders as well as vigorous partnerships on projects of mutual interest. In this section, conditions fostering collaboration are discussed, obstacles to collaboration are identified, and possible ways to overcome those obstacles are suggested.

Communication, both talking and listening, is key. WRD needs to inform the public of planned, ongoing, and completed projects and their products (reports, maps, Web pages, data, fact sheets, etc.). WRD should not presume public awareness or support of its efforts, so communication must be given high priority, and efforts to improve communication must be persistent. Education and outreach are recognized avenues of public communication, but "communication" implies an exchange of information and a relationship with the public. Through education and outreach, WRD can capture public attention and work *with* the public to help define public value. The public in the broadest sense, as stakeholders, can help identify valuable activities or products. WRD should not presume what the public values and should continuously seek new ways of creating public value. Effective listening is a matter of maintaining a receptive, service-oriented stance. Channels will open with vigorous partnerships with public groups such as watershed groups or, if necessary, advisory/liaison committees representing a cross section of stakeholders should be established. Partnerships are preferable in that they tend to be more interactive.

The experience of this NRC committee suggests that obstacles to collaboration are both institutional and practical. Institutionally, WRD employees do not always feel authorized and encouraged to start new programs with unconventional structures. Practically, many WRD employees do not have much contact with potential collaborators. Both obstacles are discussed in this section.

Does WRD have the authority to undertake programmatic innovation? Although the agency has an established mission, flexibility exists. It is helpful to view WRD employees as public managers who are expected to create public value (Moore, 1995); WRD's activities ultimately are authorized and sustained by citizen perception that the agency is creating something of value. Legitimizing WRD activities requires that the agency be attuned to the nation's needs and that citizens be aware of the value created by WRD. The interface between WRD and citizens is the production and use of reports, streamflow and water quality data, maps, and fact sheets. That interface should be broadened and should be made as dynamic as possible through ceaseless attention to improving products and their availability (Chapter 5 discusses the critical issue of access to data). WRD needs to engage the public in a two-way discourse as to what is needed and valued. Where this discourse has developed and WRD is working collaboratively, stakeholders at various levels assume an advisory as well as a client role with respect to WRD.

Resources for the GWRP have been shrinking, as discussed later, but WRD products and services are sufficiently valued that additional resources can be brought to bear by an enterprising district office. District chiefs should feel authorized to seek new activities that create public value and that are sustainable; close association with the public via multipartner collaborations, combined with an eye to the WRD's mission, will ensure that the agency is attuned to value and is aligned with its own strategic plans. To encourage innovation, the director of the USGS should support and reward innovation when it occurs.

As a practical matter, collaboration in multipartner efforts of great public value requires that partners physically meet—in the field, at seminars, and above all around the conference table. Proximity of WRD offices to potential partners has fostered collaboration in a number of cases. For example, the WRD district office in Madison, Wisconsin, shared a building with the state geological survey, and both were located adjacent to the University of Wisconsin-Madison campus and not far from other state agencies. As a consequence, collaboration among these groups started early and has survived the relocation of several of the partners. This does not necessarily recommend office proximity, but it does recognize that collaboration is a practical matter of individuals meeting face-to-face. WRD may if necessary take the initiative and follow the Middle Rio Grande basin model, hosting meetings and workshops, publishing abstracts or short papers prepared among partners, and maintaining Web sites for large collaborative projects.

Collaboration should not be imposed as a requirement on WRD scientists; instead, incentives should be offered and institutional barriers removed to permit collaboration. WRD scientists should be allowed discretion to solicit non-WRD scientists as collaborators, and the organization should consider making a general solicitation for non-WRD scientists to work on agency projects. WRD should encourage collaboration on projects funded by other agencies such as the National Science Foundation (NSF), EPA, National Institutes of Health (NIH), and DOE, which can also provide a means of increasing the funding base for fundamental or more research-specific activities.

INTERNAL COLLABORATION

Collaboration with other programs *within* the USGS is also a necessity, rather than an option, for the GWRP. The annual budget for the

GWRP (and its predecessor, RASA) has declined from more than $15 million in the early 1980s to $3.1 million in 1999 (Figure 2.2). The GWRP presently accounts for less than 2 percent of the budget of the WRD (Figure 3.2). To put the cost of this program in perspective, it is instructive to compare the GWRP budget with budgets for groundwater mitigation. The Comprehensive Environmental Response, Compensation, and Liability Act (CERCLA) of 1980 ("Superfund") was established to clean up hazardous waste sites that threaten human health or the environment. CERCLA and the Superfund Amendments and Reauthorization Act (SARA) of 1986 authorized over $15 billion through 1997 for cleanup (Guerrero, 1999).

Despite this disparity of expenditure, the framework provided by groundwater assessment and science often underlies successful efforts at mitigating contamination, as well as many other groundwater management issues of national significance. The GWRP is a small program with a large mission. Clearly the funding is not sufficient to conduct a national, process-oriented, multiscale synthesis of the nation's groundwater resources, but large budget increases for the WRD are unlikely. Financial constraints make it essential to maximize the value of data and information generated by every issue-specific or aquifer-specific study conducted by the WRD. These constraints also raise the question of whether some of the regional studies might be conducted under the auspices of one of the other WRD programs.

At first glance, internal cost sharing within WRD appears unlikely. Existing programs have well-defined missions, and resources are fully committed. The National Water-Quality Assessment (NAWQA) Program focuses strictly on water quality. The National Research Program and the Toxic Substances Hydrology Program focus on important but rather specific inquiries that usually lack a regional scope. The Hydrologic Networks Program is largely confined to surface water flow measurement. However, various cooperative mechanisms are available, as described in the following sections.

Federal-State Cooperative Water Program

The Federal–State Cooperative Water (Coop) Program itself is predicated on interagency collaboration and integration. The WRD shares costs (up to 50 percent) with state and local agencies on priority

Institutional Integration and Collaboration 59

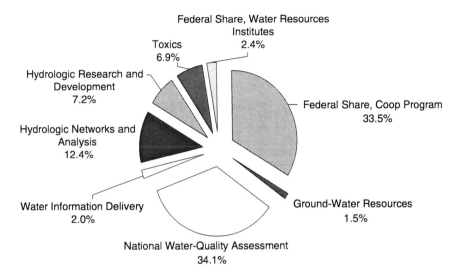

FIGURE 3.2 Ground-Water Resources Program budget as a percentage of total USGS water resources programs for FY 1999. Note that the GWRP currently represents less than 2 percent of the overall WRD effort. SOURCE: USGS.

water resources projects, with about half the funds being earmarked for data collection and half for investigations and research. The number of cooperating agencies has grown from 697 in 1982 to 1,238 in 1997, with much of the increase coming from cooperative projects with local governments.

In some projects the WRD plays the major role, providing most if not all the expertise to the project. Examples of this include the development of a numerical groundwater flow model of the Albuquerque basin for the city of Albuquerque, New Mexico (Kernodle et al., 1995), and the *Water-Resources Reconnaissance* series publications of the Nevada Division of Water Resources, which characterized and inventoried the water resources of that state. Cooperative work can lead to projects in which there is more true collaboration, as other organizations assume the roles of partners in performing the work. The Middle Rio Grande basin study in New Mexico, which evolved from a Coop Program modeling study into a partnership involving numerous agencies and other entities, is a prime example.

The Coop Program and others like it (e.g., DODEC) have also provided models for cooperation *within* the WRD, by providing important feedback to WRD research and methods development (NRC, 1996). For example, when local Coop projects gradually exposed the widespread nature of contaminants in groundwater in the 1970s, the WRD responded by developing new procedures for sample collection, new analytical methodologies, and new approaches to understanding the transport of organic contaminants in groundwater (NRC, 1996). This emphasis on organic contaminants helped launch the WRD's Toxics Substances Hydrology Program.

With coordinated effort, many activities of the Coop program can and should be aligned with broader water resources objectives to achieve a regional synthesis. As previously described by this committee (NRC, 1994):

> The development of hypotheses concerning the extrapolation of findings to other areas is crucial to establishing national relevance. The primary role of regional synoptic studies will be to test hypotheses about broad regional patterns based on inferences from study unit investigations and existing data. This iterative process of developing and testing hypotheses about large-scale regional patterns from small-scale studies of processes, and, conversely about processes and other influences from large-scales studies of patterns and trends, is essential to the synthesis process. This approach to national assessment is essentially a stratified study design, in which the strata of conditions may not be definable at the outset but can be developed as new information is added.

USGS district chiefs can be proactive in selecting projects in line with regional needs because while Coop research agendas are driven locally, resources are limited so the USGS has latitude in selecting projects that address a wider need. Likewise, districts may be able to regionalize a local study by bringing partners from adjacent political units (e.g., counties) to the table. One test of the success with which the strategic objectives of regional and national synthesis have been successfully integrated into local cooperative studies may be the extent to which individual study managers have identified the feedback and relationship between the critical issues and drivers in their specific study areas, and the extent to which they have identified regional and national issues common to other place-based studies in the National Research Program.

An added incentive exists for district chiefs to align district research with national priorities: with an expanding private sector skilled in hydrogeology and a sometimes narrowing focus of the WRD to smaller-scale Coop studies, the potential for competition between the WRD and environmental consultants has developed. The environmental consulting industry expanded in the 1980s and 1990s, in response to the enactment of CERCLA and the Resource Conservation and Recovery Act (RCRA). In 1960, 65 percent of 3,000 hydrologists surveyed worked for the federal government, while 10 percent were employed in the private sector (NRC, 1991b). By 1988, these figures were 30 percent and 32 percent, respectively. During this time, moreover, the ratio of surface water to groundwater hydrologists decreased from 2:1 to 0.7:1 (NRC, 1991b).

Conflicts have arisen. The American Institute of Professional Geologists (AIPG) formally protested several USGS WRD projects in northern Arizona because the AIPG felt that the WRD had "marketed their services to and obtained projects from local entities, at the expense of private sector companies" (Garcia, 1998, p. 9). The conflict was eventually resolved, but this incident illustrates the increasingly fine line the USGS must follow in undertaking projects for the public good while avoiding competition with the private sector. On the other hand, the committee feels it is only fair to point out that without the sustained effort by USGS to develop the technological base used by consultants and others, the private sector would not have been as successful. Balance and perspective are needed. The NRC stated a decade ago that "in some cases, the WRD is acting as a consultant to local government; this situation should be avoided unless some broader purpose is served. The WRD should continually evaluate the merits of its local assessments and cooperative activities to ensure that its limited personnel are engaged in projects with a scientific or national purpose" (NRC, 1991a). We believe that this statement still holds.

National Water-Quality Assessment Program

The National Water-Quality Assessment (NAWQA) Program, initiated in 1991 (Leahy and Wilber, 1991), is often mentioned as a model for the proposed regional groundwater assessment and national synthesis. NAWQA is designed to describe the status and trends in the surface and groundwater quality of a large portion of the nation and identify cause-and-effect relationships between water quality and human/natural factors. The program also has a national synthesis component (NRC, 1994), which is designed to "scale up" information obtained from the 60 study units into a coherent assessment.

Like the NAWQA national synthesis, the synthesis of regional and national groundwater investigations should not simply be a compilation of information from individual study units. Rather, synthesis must use the information and understanding obtained through the collective efforts among most or all of the individual study units to obtain a broader, process-oriented understanding that can be used to support program decisions and policy-making at the national level. There is an inherent trade-off, and a delicate balance must be maintained between efforts directed toward local issues and issues perceived to be of national significance. The need for balance is critical for a program like the GWRP; leveraging limited resources in cooperative studies is the only opportunity for supporting a consistent national synthesis.

The NRC (1994) recommendations on the need for a process-oriented framework for NAWQA synthesis are also appropriate for regional groundwater studies and are reiterated here. Essential elements that should be integrated throughout the Ground Water Research Program include coordination and common linkages between data collection efforts (in a variety of settings and contexts) and process-oriented research, uniform protocols for sampling analysis and archiving of data (including model parameterization and source code), and information structures that allow information flow between local cooperative studies and regional and national synthesis.

Just as the NAWQA Program was initiated and refined through pilot studies in seven different regions (e.g., the Carson River basin, the Delmarva Peninsula, and the Central Oklahoma Aquifer), the USGS's current regional assessments in the Middle Rio Grande basin can be viewed as a prototype for the development of consistent protocols for regional and national assessment.

The NAWQA Program utilizes two types of committees to support its programs. The first type of committee is a single Federal Advisory Council (FAC), comprised of federal agency representatives. This council meets every six months. Its duties (NRC, 1994) include the following:

- assist in the selection of study units and national synthesis topics,
- exchange and discuss assessment protocols, and
- discuss and review assessment findings.

The second, a liaison committee, was formed for each study unit. Membership consists of representatives from organizations noted in Table 3-1. As these committees were formed, the WRD sought a balance of management, research, and regulatory interests. These committees meet about every six months and provide important local and regional input, specifically performing the following functions (NRC, 1994):

- exchanging information about water quality issues of local and regional interest,
- identifying sources of relevant data and information,
- discussing adjustments in project design,
- collaborating on the collection and analysis of data and information,
- assisting in the design of information products from the projects, and
- reviewing and commenting upon planning documents and project reports.

Much of this structure could be adapted for the regionalization of the GWRP.

Toxic Substances Hydrology Program

Another example of interaction is the Toxic Substances Hydrology (Toxics) Program, a program formally initiated in 1983 but existing previously as the Subsurface Waste Injection Program. In 1986, a surface water component was added. The program's objective is to provide information that will be useful to decision-makers in remediating existing waste and preventing future contamination. Intensive field investigations at contaminated sites are designed to (1) obtain a better understanding of the processes controlling contaminant fate and transport, (2) gather information that is transferable to other sites, and (3) address the major sources and types of contamination in groundwater and surface water (NRC, 1996). Early in the program the decision was made to conduct long-term research at well-characterized sites, and projects are funded as long as they are productive (NRC, 1996).

The Toxics Program has several features that would be valuable elements in a regional groundwater program. First, the emphasis on *long-term* studies at well-characterized sites has yielded important insights into processes. Second, transferability has been emphasized. Third, the program has provided fertile ground for collaboration.

Although collaboration is not a primary concern of the Toxics Program, its emphasis on long-term site-specific studies has attracted researchers from outside the USGS, most notably from the academic community. One of the best examples of this is the work at the Massachusetts Military Reservation on Cape Cod (http://mass1.er.usgs.gov/-CapeCodToxics/). One of the studies involves the fate and transport of chemicals in a contaminant plume from a sewage treatment facility of the now-decommissioned Otis Air Force Base. WRD personnel performed initial work, but later the WRD invited university researchers to conduct their own experiments (e.g., Krueger et al., 1998). The long-term nature of the investigation, the well-defined site hydrogeology, and the site's security have attracted many researchers and graduate students from regional universities (e.g., MIT, Yale, and University of Waterloo). A similar but smaller-scale example is a site of an oil pipeline spill in Bemidji, Minnesota, where university and WRD researchers have made important findings vis-à-vis natural attenuation of organic contaminants.

The Toxics Program has also produced cooperation with other federal agencies. In 1989 the program began studies on the occurrence of agricultural chemicals—pesticides and nitrate—in the waters of the upper Midwest corn belt states. These studies have been coordinated with the EPA and the USDA (NRC, 1996).

CONCLUSIONS

The complex and multidisciplinary nature of groundwater resource problems of regional and national importance dictates the need for greater collaboration of expertise within the WRD and with other federal agencies, state and local governments, and the private sector. Project leaders and other WRD scientists should also be encouraged to collaborate with researchers from outside the USGS. They should also fully

Institutional Integration and Collaboration 65

participate in local, regional, and national conferences, technical meetings, and workshops to examine the larger scientific and societal context of their work.

Within the WRD, greater synergy can be achieved by aligning as many activities of the Coop and other programs as possible with regional water resources objectives. Districts may be able to regionalize a local study by bringing partners from adjacent political units to the table. Regional directors may assist in the coordination of synoptic studies by removing institutional obstacles to collaboration and by encouraging programmatic innovation. This regionalization of local assessments and cooperative activities will also help to avoid conflicts with the private sector.

Uniformity in protocol in process-oriented groundwater research and in data collection, analysis, and archiving (including model parameterization and source code) should be sought whenever possible. Information should be structured to flow easily from local cooperative studies to a regional and national synthesis. Existing models for information management, such as those used in the NAWQA and Middle Rio Grande basin programs, should be examined as potential prototypes for regional assessments.

Technical advisory committees—consisting of water managers and planners, university researchers, representatives from local, state, and federal agencies and from citizens' and environmental groups, and other stakeholders—should be established for regional study units. Advisory committees assist in the selection of study boundaries, develop assessment protocols, convey local and regional values and interests, collaborate on the collection and analysis of existing and new data, assist in the design of information products, and review planning documents and project reports. The active participation of the public not only as clients for information, but also as advisors, will help the WRD to obtain stakeholder allegiance and support for new and existing activities.

4

Scientific Issues

In Chapter 1, some of the nation's most pressing groundwater issues, along with their social importance, were introduced. This chapter presents most of the same issues, with their corresponding tools or methods, as potential research topics for incorporation into the Ground-Water Resources Program (GWRP), and provides recommended actions for the USGS. The issues are the following:

- aquifer management,
- natural groundwater recharge,
- groundwater quality and movement in surficial materials,
- groundwater–surface water interactions,
- groundwater in karst and fractured aquifers,
- characterization of subsurface heterogeneity,
- modeling of flow, transport, and management, and
- facilitating the use of groundwater information in decision-making.

One common thread that connects all the topics discussed below is the necessity of integrating geochemical investigations into many, if not most, groundwater studies. The committee recognizes that most groundwater problems have a significant geochemical component and that geochemistry can often provide important insights into hydrogeologic processes. Historically, most regional groundwater investigations by the USGS have emphasized physical hydrogeology at the expense of geochemical hydrogeology. Yet physical and geochemical problems are usually intertwined, and both affect sustainability.

AQUIFER MANAGEMENT

Scientific and Management Issues

Water managers have the very real problem of trying to project water use and water supply for a future that includes population growth, climate variability, and unknown technological breakthroughs. They must make decisions about curbing growth, investing in technology, and balancing the various needs of stakeholders and ecosystems. Fundamental to these decisions are water-budget issues. How much water can be used without drawing down the water table or potentiometric surface, thereby causing loss of storage, salt-water intrusion, or property damage due to aquifer subsidence? How can managers avoid drying up streams or draining wetlands—many of which retain suspended sediment, excess nutrients, and pesticides—and maintain wildlife habitat? The sustainability of human communities, including the ecosystems that support them, needs to be considered as an integral part of aquifer management. Climate change over decades can also have a major effect on water resources, independent of local human influence. Changes in global weather patterns can cause marked changes in precipitation and evapotranspiration rates and distribution, resulting in changes in recharge, streamflow, flooding, and drought patterns. Although fully understanding climate change is a global issue, the USGS has a useful role in assisting those predicting climate change at the regional level in the United States.

Excessive pumping of groundwater for irrigation and other uses has caused water-level declines of greater than 100 feet in some regions. In addition to causing resource depletion, this reduces pore pressures and raises the effective stress on the aquifer, often leading to irreversible consolidation. Differential settling cracks foundations, which may not only be costly for structures such as roads or buildings, but also hazardous to dams, power plants, or pipelines. In some cases, subsidence caused by irrigation pumping has lowered the land surface to the point where rivers have changed course and have flooded agricultural lands. Subsidence caused by overpumping of both water and hydrocarbons has submerged coastal areas below sea level, causing ecological damage to coastal wetlands and exacerbating hurricane damage (White et al., 1993; Kreitler, 1977).

Excessive pumping has caused salt-water intrusion in the majority of U.S. coastal states, including Massachusetts (Person et al., 1998), New Jersey (Pope and Gordon, 1999), South Carolina (Smith, 1994), Florida (Merritt, 1996), Louisiana (Tomaszewski, 1996), and California (Izbicki, 1996). Even inland areas underlain by formations containing saline water are susceptible (Sophocleous and Ma, 1998). Saline groundwater is present in most of the major basins of the United States, and as coastal cities grow, this problem may be expected to get worse.

It may take decades before salinity is noticeable in well water and, by then, years may be needed to purge the saline plume even if pumping halts. Injection of freshwater hastens the purging process only slightly, assuming a fresh supply can be found (Kazmann, 1972). For this reason, it is in the national interest to thoroughly understand the process of salt-water intrusion to assess and manage the risk before damage occurs.

The position of the freshwater–saltwater contact can often be estimated using the Ghyben-Herzberg principle (Baydon-Ghyben, 1888), which predicts a density-controlled floating lens of freshwater with a "root" approximately 40 times the elevation of the water table above sea level, thinning toward the coastline. However, most salt-water intrusion problems are too complex for simplistic approaches. Wedges of relict freshwater occur far offshore, sandwiched between saline water. Tidal forcing, rainfall events, and storm surges are transient short-term processes that influence salinity. Pockets of relict seawater and intrusion of salt water through failed well casings, joints, or sinkholes complicate the interpretation of salinity in well water. Heterogeneity in aquifer material properties also affects the location of the saltwater–freshwater interface. Such systems are generally studied today using either a sharp interface model (e.g., SHARP; Essaid, 1990) or a variable-density solute transport model (e.g., SUTRA; Voss, 1984).

In addition to modeling, methods used to investigate freshwater–saltwater interactions include tracers (Box 4.1) and geophysical tools, especially electrical methods that are sensitive to conductive water (Rozycki, 1996). Uncertainty in and scale dependence of material properties and processes plague virtually all measurements of the saltwater–groundwater flux.

Aquifer storage and recovery (ASR) technology has recently gained

Box 4.1
Examining Freshwater–Saltwater Processes with Radium Isotopes

Global biogeochemical cycles are affected by the complex exchange of fluvial, subsurface, and marine material in estuaries. The accurate quantification of these fluxes is no simple task. Satellite imagery can be helpful but cannot generally provide information on the movement and rate of water mixing. Natural and artificially produced radioisotopes can be used as tracers to ascertain mixing and exchange of material across the sediment–water interface. The USGS is using four naturally occurring radium isotopes (radium-223, -224, -226, and -228), with half-lives ranging from 3.7 days to 1,600 years, as tracers (Swarzenski, 1999). Radium isotopes are part of the uranium-238, thorium-232, and uranium-235 decay series; they are produced by decay of thorium isotopes.

In freshwater, radium is bound to sediment particles; as the water becomes more saline, the radium moves to the dissolved phase, such that in the open ocean, it resides exclusively in the dissolved phase. In contrast, thorium remains bound to sediment particles. Thus, estuarine sediments provide a continuous supply of radium isotopes to coastal waters. The combined source functions for radium in an estuary include (1) riverine particulates and dissolved, (2) oceanic dissolved, (3) estuarine sediments, and (4) groundwater. The relative significance of these sources is a function of the site-specific hydrogeology and of where the samples are taken relative to the salinity gradient (determined by the extent of freshwater–saltwater mixing). Because groundwater is commonly enriched in radium relative to surface water, a transient groundwater influence can be seen, even in a dynamic water column. In surface sediments that are flushed with groundwater, a disequilibrium develops between radium-228 and thorium-228, which permits an estimation of groundwater fluxes or an apparent age of the water mass.

The USGS is currently applying this technique to Florida Bay, which receives much of its freshwater from Taylor Slough, a wetland along the northeast boundary of the Everglades. The bay is vulnerable to deteriorating water quality in the Everglades, and it may also be receiving anthropogenic contaminants through subsurface flow. USGS scientists are currently developing models to constrain groundwater flux estimates.

increased interest as a means for aquifer management. Aquifer storage and recovery projects involve the artificial storage of water in underground aquifers during times of water availability and the recovery of that water when the water is needed (Pyne, 1995). Most projects involve the subsurface injection of water into aquifers and later extraction of the same water. Example ASR applications to meet aquifer management needs include, among others, seasonal storage of water, emergency storage of water, the prevention of salt-water intrusion, enhanced wellfield production, and hydraulic control of contaminant plumes. ASR technology has been used in various parts of the nation since the late 1960s. The use of ASR poses many technical challenges. These include assessing the hydraulic performance of the systems and determining the effects on nearby wells, the long-term geochemical changes caused by mixing waters of different chemical compositions in the subsurface, and contaminant migration away from ASR sites.

USGS Roles in Aquifer Management

The role of the USGS in aquifer management includes collecting, inventorying, and analyzing data on groundwater levels, developing improved techniques for acquiring such data, and developing and improving analytical and numerical tools for aquifer management.

Potentiometric and water-level maps are a key tool in assessing the effects of regional water use. Such maps were made for the major regional aquifer systems as part of the Regional Aquifer-System Analysis (RASA) Program. An unmet need is a national effort to track water levels over time in order to monitor water-level declines (Sun and Johnston, 1994). This is being done on an ad hoc basis by individual states, but the creation of *regional* potentiometric maps is the responsibility of the federal government. Data to support potentiometric surface mapping are likely to be available from non-USGS entities, especially state geological surveys; the USGS must collaborate with these entities in sharing and interpreting water-level data.

Traditional groundwater resources projects are still required in many areas. A growing U.S. population, especially in the arid and semiarid regions, points to the need for the USGS to explore and characterize alternative groundwater supplies in areas such as the Great Basin in Nevada.

Scientific Issues

The exact location and rates of subsidence depend on geology and duration of pumping. Although it is not possible to make exact predictions with respect to settlement, it should be possible to improve our performance in that area. More effective management will first require better definition of geologic heterogeneity. The USGS should continue to study the relationship between water levels and subsidence, modeling interactions among lithology, clay content, recharge, pumping, storage, and subsidence. The goal should be to keep subsidence within safety limits for the strain of structures and to identify the critical pumping rate at which there is no permanent strain.

For tracking regional subsidence, techniques such as synthetic aperture radar interferometry (Massonnet and Feigl, 1998; Amelung et al., 1999) and global positioning systems (GPS) should be fully exploited. These do not require a fixed datum, as does high-precision leveling, and are more cost-effective for large geographic areas. Likewise, arrays of piezometers with transducers should be used to track long-term regional changes in the potentiometric surface as an early warning system. Borehole tilt-meters or seismographs can be deployed in high-risk areas.

The NRC identified the need to analyze links between water resources and climate change as one of eight key areas for USGS WRD research (NRC, 1991a). The difficulty of scaling hydrologic models to be compatible with coarse-meshed global circulation models, or vice versa, is a limitation that must be overcome. Predictions under a variety of scenarios must be wedded to decision-making models—they must be presented in a form useful to water managers and decision-makers.

Salt-water intrusion modeling has far to go before it reaches the stage where it can be used effectively by water resources managers. Surprisingly, few test cases exist for independently "verifying" the groundwater codes used for such modeling (Simmons et al., 1999). Also, although three-dimensional models exist, computation time still limits most real-world simulations to two-dimensional analysis. A further challenge is that the nonlinear coupling of the flow and transport equations creates difficulties in their numerical solution. Finally, integrated optimization tools are generally lacking, as are linkages to geographic information systems (GIS).

Geochemical methods can also use refinement. The radium tracer technique of Box 4.1 has potential for field ground-truthing. Various ratios—e.g., Cl:Br (Davis et al., 1998) and Cl:F (Vengosh and Pankratov, 1998)—have also shown promise in distinguishing Cl from modern seawater, "connate" water, wastewater, road salt, and domestic water-conditioning recharge effluent.

It is likely that the USGS will become involved with ASR projects as they influence regional aquifer management. Appropriate roles for the USGS include regional modeling of ASR impacts, investigations of geochemical and hydraulic processes associated with ASR projects, and determination of aquifer properties (transmissivity, storage, heterogeneity) relevant to ASR performance and design.

NATURAL GROUNDWATER RECHARGE

Groundwater recharge is a critical part of the water budget, and it is arguably the hardest component to quantify. The difficulty in measuring this "income" term in the water budget makes it no less important, especially in arid and semiarid areas. It is also important in coastal areas, where lowering of the water table induces salt-water intrusion into water supplies, and in surficial aquifers, where recharge can carry surface and soil contamination into shallow water supplies.

Scientific and Management Issues

The critical attributes of recharge are its rate and spatial distribution. Combining the two yields volumetric recharge to an aquifer. In the past, the estimation of recharge *rate*, particularly in arid areas, was given the most attention. Recharge rates can be estimated using hydroclimatological approaches—requiring measurement or estimation of rainfall, evapotranspiration, soil moisture, and runoff—and treating recharge as the residual.

Groundwater recharge rates and their spatial distribution can also be estimated using environmental tracers such as dyes, chloride, bromide, nitrogen-15, and chlorofluorocarbons (CFCs), the stable isotopes deuterium and oxygen-18, and the radioisotopes carbon-14, tritium and chlorine-36 (Clark and Fritz, 1997, pp. 80–99). These techniques can be applied on both the small scale (Cook et al., 1994) and large scale (Campana and Boyer, 1996). Agricultural chemicals for which application records exist can also serve as tracers.

Recharge can also be estimated by intensive study of infiltration and moisture redistribution in the unsaturated zone. Obviously, the scales of these two approaches are drastically different, from tens of kilometers to centimeters. The infiltration approach is physically based and rigorous; however, extrapolation to large scale presents an obstacle. Moreover, although the centimeter-scale process can be modeled using physically based models of the unsaturated zone, linking these models to aquifer models with a resolution of tens of meters or kilometers continues to be difficult. The lack of data to support centimeter-scale modeling of vast areas provides a strong disincentive to reconciling the two scales.

Methods of measuring recharge directly have the advantage of integrating the sub-centimeter-scale changes to the meter scale. Although the controlling processes occur at the pore scale, they result in a perceptible movement of moisture that can be measured at the field scale with appropriate field instruments. Noninvasive surficial methods include geophysical methods and remote sensing. Time-domain reflectometry and micro-gravity surveys show promise in determining recharge rates (e.g., Young et al., 1997). The USGS is monitoring micro-gravity at the University of Arizona's network near Tucson in the first basinwide application of micro-gravity methods to the measurement of changes in groundwater storage. Long-term monitoring, including two El Niño events already, will permit correlation of storage changes with climatic events, facilitating water-use planning and management. Recharge has been interpreted from remotely sensed data with some success. For example, high-resolution radar images, filtered by principal components analysis, show promise for quantifying the dependence of recharge on climate and topography (Verhoest et al., 1998).

Although the USGS and others have been researching various methods of estimating recharge, the goal of straightforward regional application has yet to be achieved in most cases.

For example, the use of ground-penetrating radar to determine travel times for establishing depth to water is confounded by variations in soil moisture. Unfortunately, surficial methods of recharge estimation will always be difficult because of spatial variability of hydrogeologic materials and soils. For example, using a water-table rise as evidence of recharge may be misleading if elevated areas are actually areas of lower hydraulic conductivity, because there is not a unique relationship among hydraulic conductivity, head, and recharge. Statistical methods of optimizing parameter estimates can be brought to bear on the problem of nonuniqueness, but describing the aquifer heterogeneity is still critical.

In most watersheds, recharge is not spatially uniform because of variations in rainfall, evapotranspiration, infiltration, and runoff. Discharge, or negative recharge, may be a natural process occurring in wetlands or stream valleys, or it may be a result of pumping. In any case, predicting the flow of water within aquifers requires specifying the fluxes of water into and out of the system. Until recently, standard practice was to assume a uniform recharge rate over an entire watershed, and for some purposes this assumption yielded practical results. The assumption becomes increasingly restrictive with decreasing scale and with increasing need for resolution. In 1991, the NRC recommended the development of methods to identify critical recharge areas on small spatial scales (NRC, 1991b). The need for mapping recharge remains, despite locally notable efforts such as Sophocleous (1992). Although methods have been proposed, they have not been widely used and are complicated by problems of scale.

USGS Roles in Groundwater Recharge

For regional studies of groundwater, it is essential that the USGS continue to develop and test methods that define recharge at scales ranging from local to regional (Box 4.2). The required knowledge base includes (1) an improved understanding of basic controlling processes such as evapotranspiration and infiltration, (2) new modeling methods integrating centimeter-scale processes and linking them to large-scale models, including numerical methods for handling nonlinearity in saturated–unsaturated models, and (3) methods to measure or average or statistically represent centimeter-scale heterogeneity.

Improved knowledge of groundwater recharge will help water managers protect aquifer health under stresses imposed by increasing withdrawals or by drought, and it will help them avoid recharging aquifers with poor-quality (contaminated or saline) water. From the point of view of health of aquifers regionally, it is critical that studies of recharge make the leap from local, intensive "case" studies to general principles, determining what controls recharge regionally and mapping those factors with a GIS to provide a basis for aquifer management. As appealing as this concept is, efforts to map groundwater vulnerability regionally for management have not always produced practical results. It is important that maps not be too generalized if they are to be useful in local man-

> **Box 4.2**
>
> **The Response of Surface Water–Groundwater Interactions to Present-Day Climate Variation—Amargosa Desert Case Study, Nevada–California**
>
> Ephemeral streams in the southwestern United States flow infrequently and are ungauged. Flow losses during runoff events produce groundwater recharge, but little is known about the timing, duration, and spatial extent of these flows. This information is needed to provide quantitative estimates of modern natural recharge and to understand the response of recharge to climate variation.
>
> The 1997–98 El Niño weather pattern provided an opportunity to investigate the effects of wet conditions on arid-land hydrology. During this time, precipitation in the Amargosa Desert was more than twice its annual average of 100 mm. In early January 1998, the USGS, led by David A. Stonestrom of the National Research Program in Menlo Park, California, placed streambed temperature sensors along 80 km of the Amargosa River. The lower 60 km of this reach are dry in normal years. Temperature data, verified by field observations, indicated that the Amargosa River flowed for its entire length, from its headwaters in Oasis Valley to its terminus in Death Valley, in February 1998. Peak-flow measurements from the temperature sensors coupled with two crest gauges (used to provide ground truth) will be used to estimate percolation losses through the streambed. Future plans call for a more extensive temperature-sensor network to provide greater resolution of percolation losses through the streambed.
>
> SOURCE: http://az.water.usgs.gov/swgwrp/Pages/Amargosa_desc.html

agement. If decisions about water or land use affect citizens preferentially, the map must be detailed enough to resolve local variations in soil, topography, and drainage perceived by an observant citizen.

GROUNDWATER QUALITY AND MOVEMENT IN SURFICIAL MATERIALS

Over broad areas of the United States, groundwater occurs in shallow surficial materials. These materials include glacial, alluvial, and lacustrine deposits as well as weathered bedrock residuum. In general, such materials are a few tens to a few hundreds of feet thick and often lie above deeper bedrock aquifers. Surficial materials can be quite discontinuous, as exemplified by eskers in the Northeast, or they can be very extensive, such as the till sheets in the northern Midwest. Where such materials are composed of permeable sand and gravel, they often form important aquifers. However, materials of lower permeability, such as clayey till or silty lacustrine deposits, also contain and transport groundwater and have important functions in the overall water cycle.

Scientific and Management Issues

Occurring near the land surface, groundwater in shallow surficial materials is particularly vulnerable to contamination (see Chapter 1, Box 1.1) by the hundreds of thousands of reported releases of gasoline from leaking underground fuel tanks nationwide, and the nation is currently spending hundreds of millions of dollars remediating contaminated sites in these materials. The USGS National Water-Quality Assessment (NAWQA) Program discovered many instances of nitrate and pesticide contamination of shallow groundwater in agricultural areas (USGS, 1999b). Likewise, onsite septic systems and lawn fertilization can also contaminate groundwater. Shallow groundwater contamination can move to adjacent lakes, rivers, and wetlands as well as to underlying deep aquifers used for water supply.

Concern for the integrity of groundwater supplies has led to legislation at all levels of government to protect aquifers from contamination by land use, much of it under wellhead protection clauses. Despite great effort expended on predicting how water and contaminants move underground, it is still difficult to state with confidence that a given land use will have a specific impact on a particular water supply. Clearly, though, regional deterioration of shallow water supplies has occurred and can be linked to land-use practices, with an example being agricultural fertilizers causing high nitrate levels in rural water supplies. Much

Scientific Issues

of the uncertainty in pinpointing sources arises from aquifer heterogeneity and complexity, confounding wellhead protection analyses.

USGS Roles in Surficial Material Hydrogeology

Knowledge of groundwater movement, quality, and quantity and of potential contaminant sources and pathways is needed to effectively manage surficial groundwater. Public-sector managers need water quality and water-level monitoring information in areas geologically susceptible to degradation or characterized by high-risk land-use practices, and they need the educational tools to enlist the support of the public in their own self-interest. Where prevention fails or contamination has already occurred, the managers need tools to restore the water supply. The USGS is providing, through NAWQA, a regional assessment of water quality. The USGS is strongly encouraged to continue developing accurate and accessible information about groundwater using available technologies.

Aquifer restoration has developed to a high degree of sophistication in the private sector, given economic incentives driven by public regulations. However, because the public interest is at stake, impartial assessments of restoration technology are needed. In particular, the cost and difficulty of aquifer restoration (e.g., difficulty in delivering nutrients and oxidants to contaminated zones for in situ bioremediation) has led to a closer look at intrinsic bioremediation in contaminated surficial aquifers (Barber, 1994; Smith et al., 1994). The process of intrinsic bioremediation, or natural attenuation, needs to be quantified before regulatory decisions can be made (Chapelle, 1999). Because the U.S. Environmental Protection Agency (EPA), other federal agencies, and universities are also active in this area, study sites should be chosen carefully for their regional applicability.

Setback requirements and other guidelines for accepted natural-attenuation systems such as septic tanks must be reviewed and revised in the face of developments in the understanding of pathogen transport and lithologic controls. The transport of microbes can be quite variable. Sorption of pathogens is microbe-specific and depends on surface chemistry and flow velocity (Harvey, 1993; Hendry et al., 1999). The USGS should be more involved in studies of natural attenuation, just as it has been involved with studies of contaminant transport at several well-documented sites.

GROUNDWATER–SURFACE WATER INTERACTIONS

Scientific and Management Issues

The USGS has long been a leader in the quantification and characterization of surface water and groundwater interactions. Winter et al. (1998) documents not just the role of the USGS, but also the evolution of knowledge in the field in general. Earlier works by Winter (1976, 1978, 1981) led to major advances in the science of groundwater–wetland and groundwater–lake interaction.

Groundwater and surface water interaction occurs in all types of hydrogeologic and climatologic settings at all spatial and temporal scales (Winter et al., 1998). Groundwater can interact with streams, lakes, estuaries, bays, wetlands, and coastal areas. The interconnection between groundwater and surface water means that they often behave as one reservoir and should be treated and managed as a single resource (Winter et al., 1998).

Groundwater, Lakes, and Streams. Early views of groundwater–stream interactions often treated streams simply as areas of recharge to groundwater or recipients of discharge from groundwater; groundwater flow paths in the vicinity of surface water bodies were generally two-dimensional and simplistic (Woessner, 1998). Interactions between groundwater and streams or lakes were often characterized in terms of base flow or bank storage, the latter being recognized as an important storage and flood-wave attenuation mechanism (Whiting and Pomeranets, 1997). Interactions in coastal regions were often cast in the context of saline-water intrusion (Todd, 1959; Davis and DeWiest, 1966; Freeze and Cherry, 1979).

Flow paths in heterogeneous aquifers near surface water bodies are far more complicated and dynamic than was previously thought (Woessner, 1998; Wroblicky et al., 1998). To the simple "gaining" and "losing" streams (and other surface water bodies) can be added "zero exchange" and, perhaps more significantly, "flow-through" streams (Woessner, 1998). Temporal changes in water and chemical exchange between groundwater and surface water occur because of variations in stream stage and discharge (Squillace, 1996; Wondzell and Swanson, 1996; Morrice et al., 1997), which sometimes occur very rapidly (Wroblicky et al., 1998).

Scientific Issues

The interface between groundwater and surface water systems is recognized as a distinct zone, the hyporheic zone (Gibert et al., 1990; Vervier et al., 1992). This refers to a region where active and dynamic exchange of water and nutrients occurs between the surface water and the adjacent groundwater system (Triska et al., 1989; Valett et al., 1997; Wroblicky et al., 1998). The hyporheic zone is an "ecotone" that allows for bi-directional flow of organisms and materials (Gibert et al., 1990; Vervier et al., 1992).

Temporal variation in the extent of exchange between a stream and associated aquifer is driven not only by the changing characteristics of the stream, but also by changes in aquifer status as well. The fluvial system includes the floodplain and hillslope where infiltration and recharge alter the hydraulic gradients linking the stream and aquifer. In this way, the variation in exchange reflects the combined effects of hydrologic change in the stream and aquifer.

Hyporheic studies, with synergistic collaboration among geologists, hydrologists, limnologists, ecologists, and geochemists, are becoming increasingly common (Triska et al., 1989; Harvey and Bencala, 1993; Stanford and Ward, 1993; Valett et al., 1996; Morrice et al., 1997; Dahm et al., 1998).

The scientific contributions of the USGS groundwater–lake interaction research program are numerous. They notably include detailed instrumentation studies designed to determine how to calculate wetland and lake water budgets and their associated measurement errors (e.g., Rosenberry et al., 1993; Winter et al., 1995; LaBaugh et al., 1997), theoretical and field investigations of wetland–groundwater hydraulics (e.g., Winter and Woo, 1990; Harte and Winter, 1995; Winter, 1995a,b; Rosenberry and Winter, 1997; Winter et al., 1998), investigation of wetland biogeochemistry related to climate change (e.g., McConnaughey et al., 1994; Schwalb et al., 1995; LaBaugh et al., 1996; Poiani et al., 1996) and, most recently, a multidisciplinary program incorporating hydrology, geochemistry, and ecology in a complex of lakes in Minnesota (Winter, 1997). The NAWQA Program, a nationwide study of surface water and groundwater quality trends and their cause-and-effect relationships, seeks to treat groundwater and surface water as a single unit, where appropriate.

The USGS also has contributed greatly to stream–groundwater interaction. Bencala and Walters (1983) formulated the concept of "transient storage" and its application to streams. Triska et al. (1989) studied the ecological consequences of groundwater and surface water exchange and its impacts on nitrogen transformations along stream corridors. Harvey and Bencala (1993) and Harvey et al. (1996) applied different hydrogeologic techniques to address how inferences on surface and groundwater interaction depend on the scale of measurement.

Groundwater and Wetlands. The National Research Program of the WRD has historically supported wetland research, its first major contributions being in the field of wetlands classification (e.g., Cowardin et al., 1979; Novitski, 1979). This research thrust has been expanded to include remote sensing and ecological research in wetlands focusing on how the distributions and kinds of submerged plants (macrophytes) may be hydrologic surrogates useful to identify water quality parameters such as salinity, turbidity, pH, nutrients, presence of various pollutants, or frequency and duration of inundation (e.g., Carter et al., 1983; Batiuk et al., 1992; Dennison et al., 1993; Carter et al., 1996).

Wetlands, as places where groundwater and surface water interact, are similar to hyporheic zones with respect to their being ecological and hydrologic interfaces between freshwater- and marine water-rich and water-poor landscapes. Compared to nonwetland areas, the unique chemical characteristics of wetland soils, their ecological community compositions, and the relative degree to which their soils are saturated vary widely across the nation's climatic regimes. The minimal essential characteristics of a wetland are (1) recurrent, sustained inundation or saturation, and (2) the presence of physical, chemical, and biological features reflective of this condition. Common diagnostic features are hydric soils and hydrophytic vegetation (NRC, 1995b). Hydric soils accumulate more organic matter than do upland soils because organic matter cannot readily decompose during the anoxic conditions that develop when wetland soils are saturated.

Determining how long and how frequently soils need to be saturated to form an ecologic wetland community is very difficult. The NRC (1995b) recommended a series of regional studies to address "how wet is wet" to make a wetland while still maintaining the overriding ecosystem-based view on wetlands characterization that typified earlier classifications. Winter (1988) and Brinson (1993) challenged the long-standing view that wetlands are fundamentally *ecosystems*. They argued instead that wetlands are *physiographic places* with hydrologic and meteorologic conditions that lead to more water, which in turn leads to a unique ecosystem evolution and unique soils over time. Brinson's hydromorphic wetland classification is largely based on water budgets and has generated wide interest in the wetland scientific community. Brinson's and Winter's classifications both fully incorporate riparian areas as wetlands.

The renewed scientific emphasis on wetland hydrology has led to fruitful research designed to understand the complex interactions among wetland ecosystems, geochemistry, and hydrology. Driving some of this research are National Science Foundation initiatives on climate change and biodiversity maintenance and the regulatory need to manage wetland resources. Wetlands are a major terrestrial sink for carbon (as peat) and are significant sources of the atmospheric greenhouse gases of carbon dioxide and methane. Wetlands are also critical landscapes that maintain marine fisheries and freshwater biodiversity (Mitsch and Gosselink, 1993; NRC, 1995b).

Research on the interactions among hydrology, geochemistry, and biology in wetlands has led to a better appreciation of solute cycling and transport. For example, we now know that (1) nutrients are processed and methane is generated by bacteria in distinct wetland soil microenvironments, best identified by sampling at the centimeter scale (e.g., Romanowicz et al., 1995; Hunt et al., 1997; Steinmann and Shotyk, 1997), (2) the extent to which this methane is generated may be related to climate change (Romanowicz et al., 1995), (3) wetland soils have significant macropores through which solutes can move by groundwater advection (e.g., Chanton et al., 1995), and (4) trace metal contents, pollen distributions, and isotopic contents of peat profiles provide important information on Holocene climate change and anthropogenic contamination (e.g., Shotyk et al., 1998).

Finally, the coupling of GIS technology with data on stream-water quality and discharge at the landscape scale has led to the conclusion that headwater depression and other small wetlands proportionally remove more suspended material and store more water than do riparian wetlands further downgradient in higher-order stream reaches (Johnston et al., 1990). This kind of research has broad implications with respect to how we manage our natural environment.

Together, the detailed studies of wetland hydrology, geochemistry, and biota and the more regional studies of wetland function in the landscape have fostered an almost entirely new scientific discipline of multidisciplinary study, complete with new journals such as *Wetlands,* published by the Society of Wetland Scientists.

USGS Roles in Groundwater–Surface Water Interactions

The central role of groundwater–surface water interactions in regional assessment suggests a number of key research areas where the USGS expertise in process-based groundwater science is essential to support the analysis and sustainable management of regional groundwater systems:

Water Availability Models. Water Availability Models (WAMs)—complex models that have their roots in conjunctive-use models—seek to simulate the flow and legal availability of surface and subsurface water in a basin, region, or political subdivision. In their most sophisticated form, WAMs also include economic, sociological, water quality, and legal models embedded within the hydrologic models. Depending on the kinds of input required, it may be appropriate for the USGS (1) to generate or apply these models directly, (2) to work in cooperation with other organizations, or (3) to be the supplier of information to third parties.

WAMs are especially useful in simulating water distribution scenarios during droughts and in managing excess water. Targeted to water rights, permitting, and planning, groundwater–surface water interactions in most water-availability models are at best represented with simplified loss coefficients, representing average estimated losses and returns (TNRCC, 1997). In many basins, regional permitting and planning decisions need to accurately incorporate the complex interactions between surface and subsurface flows in water-availability modeling. This growing need represents a timely opportunity for the USGS to integrate its expertise in process-based regional groundwater science with regional assessments of sustainable surface water and groundwater resources.

Groundwater and the ecology of surface systems. The USGS has been one of the leaders in the important area of groundwater and the ecology of surface systems. The influence of the physical and geochemical characteristics of the coupled surface–subsurface hydrologic system on the associated ecosystems, and vice versa, is not well understood. Organisms in the hyporheic zone may be important indicators of ecosystem health; this avenue merits further exploration. The USGS is encouraged to engage in integrated physical, geochemical, and isotopic studies of these interactions.

Watershed planning and management. Principles of groundwater–surface water interactions must be integrated into the increasingly common watershed-based approach to planning and management. Groundwater flow systems will often not coincide with watershed boundaries, adding to the complexity in evaluating flow paths, source areas, and dominant processes controlling streamflow, contaminant transport, and nonpoint pollution. Storm flows can be dominated by subsurface flow (Barnes, 1939; Anderson and Burt, 1980) with the chemical signature of "old water" (Sklash and Farvolden, 1979; Rice and Hornberger, 1998). The origins and transport of pesticides in streamflow are not easily resolved without detailed understanding of the response time and of interactions between direct surface runoff, bank storage, and subsurface flow (Squillace et al., 1993). Groundwater–surface water interactions embody critical processes, essential for understanding basin hydrology and watershed-scale fate and transport, developing watershed-scale management plans, and calculating total maximum daily loads.

GROUNDWATER IN KARST AND FRACTURED AQUIFERS

Scientific and Management Issues

Unlike many of the nation's major aquifers that owe their productivity to primary (i.e., intergranular) porosity, "karst" aquifers such as the Floridan and Edwards aquifers are characterized by secondary porosity due to dissolution, usually of carbonate rock, along pores and fractures. Fractures in less-soluble, low-permeability rocks such as granite, gneiss, shale, and clayey till may also yield aquifers that are relatively poor but regionally important because they are a region's only water supply. These karst and fractured aquifers present a variety of challenges to quantitative hydrogeology.

Karst Aquifers. Of all geologic settings, karst may be the most susceptible to groundwater contamination because water infiltrates rapidly, the aquifer has poor filtering ability, and residence times are short (Figure 4.1). Many regions of the United States, especially the midwestern and southeastern states, have groundwater problems associated with karst. Flow velocities can exceed 1,000 m/day, and dissolved substances, microbes (*E. coli* and *Giardia* cysts), sewage or manure parti-

FIGURE 4.1 Greer Spring in the Missouri Ozarks is the largest spring on U.S. Forest Service land, and averages 187,000,000 gallons per day (courtesy of Randy Orndorff, USGS).

cles, and even crustaceans can move through conduits in karst; apertures are sufficiently large that sediment can be transported with adsorbed toxic metals or organic molecules. The use of karst sinkholes for rural waste disposal puts water supplies in rural areas at special risk. In regulating septic setbacks, it is not clear that ambient mechanisms in fractured or karst aquifers can meet regulatory requirements (Chapelle, 1999).

Effective management means protecting the actual source area for potential water supplies, as is done in Florida by creation of conservancy recharge parks. Mapping recharge areas, however, is not straightforward. Recharge areas are controlled by fractures that create anisotropy—a strong preferred flow direction elongates the source area for a well or spring. Solutions widen certain fractures to create conduits, the locations of which are hard to determine. In fact, our ability to represent the architecture of actual karst systems is far from satisfactory.

Although conceptual models of flow in karst areas have been constructed (Bögli, 1980), numerical modeling of karst has either treated the conduits as discrete fractures, assuming no matrix flow, or has assumed the whole system to behave as a "black box." In the latter approach, recharge is the input and spring flow is the output. Modeling of karst aquifers as equivalent porous media may be feasible if the cell size is sufficiently large, and such a model provides general information on flow direction. However, head, permeability, and water quality measured in wells cannot be expected to be representative. Traditional tools of the hydrogeologist often yield ambiguous data. Placement of monitoring wells is critical, and interpretation of pumping tests, head values, and geochemistry is difficult because the concept of the water table breaks down in this highly transient, dual-porosity system. Methods need to be developed to either represent the true aquifer complexity or, if this proves infeasible, to sample the aquifer parameters in some integrated fashion.

Fractured-Rock Aquifers. Fractures in otherwise impermeable rock present some of the same problems as karst conduits. The bulk hydraulic conductivity of a fractured aquifer may be low, but flow velocities and contaminant transport rates in fractures or solution cavities can be high, resulting in early contaminant appearance at target wells or streams and in poor filtration of pathogens. Locations of fracture zones are extremely important to the development of water supplies in areas with generally unproductive aquifers. Fractured zones may be concentrated along lithologic contacts or near the surface in the case of stress-relief fractures. They may cluster around regional fault traces. Prediction of the occurrence of fracture zones based on tectonic or structural models would assist in water-supply development. Knowledge of fracture-flow dynamics and fracture orientation and location is critical because fractures may cause markedly anisotropic behavior, especially in contaminant transport.

There has been an evolution in the understanding of basic processes in fractured aquifers paralleled by developments in the mathematical simulation of fractured systems, and much of this work has been conducted by USGS scientists (e.g., Shapiro and Hsieh, 1996). Early models represented fractures as parallel plates of varying aperture, but recent models accommodate two- or three-dimensional networks of fractures with statistically distributed apertures and spatial correlation of aperture width. Representation of fractured-aquifer porosity has evolved from fracture porosity to a dual-porosity representation including matrix porosity, and even a fracture-skin porosity (Robinson et al., 1998). Diffusion, adsorption, and degradation of contaminants in fractured aquifers can be modeled. Realistic-looking fracture networks based on a growing number of field studies of fractured aquifers can be modeled, although there is still a need for a geostatistical description of fractured systems.

Modeling continues to pose many challenges, the greatest of which is the data deficiency relative to the needs of a good predictive fracture-flow model. Practical conceptual models of fractured systems are still needed. For example, in a sedimentary sequence in New Jersey, shallow flow was found to occur via bedding-plane partings, but deeper flow occurred via high-angle fractures (Morin et al., 1997). We still know little about the control of the compressive-tensile stress regime on the fractured-flow system, and there is an accompanying data need for sound management models of fractured systems.

Determining reactions along groundwater flow paths requires a sound conceptualization of the hydrogeochemical system, including velocities, residence times, surface area, oxidation state, availability of particles such as clay and organic matter, and microbes. Advection dominates transport in fractures; diffusion dominates in blocks (Moench, 1995). Contaminant pathways may be complex, irrespective of averaged heads and gradients. Natural attenuation depends on the complex reactions expected in these highly buffered systems, including dilution, redox reactions, dissolution, precipitation, adsorption/desorption, complexation, and ion exchange.

USGS Roles in Karst and Fractured-Rock Studies

Representing the architecture of a small number of actual karst or fractured-flow systems via intensive studies is a prerequisite to understanding these systems. Such studies should aim at developing methods to either represent the true aquifer complexity (perhaps as a geostatistical description of fractured systems) or to sample the aquifer parameters in some integrated fashion. The USGS Geologic Division should be recruited to integrate tectonic and structural models with hydrologic models, in order to predict the occurrence of karst or fracture zones. In particular, how does the compressive-tensile stress regime relate to the fractured flow system? Neotectonically active zones are primary candidates for preferential flow paths in karst, for example. A sound conceptualization of the hydrogeochemical system in fractured or karst regions is needed as well. Specifically, research on the fate and transport of nutrients and microbes in karst aquifers is urgently needed.

Drilling, sampling, and flow-determination protocols need to be tested and developed. For example, in drilling, borehole location relative to fractures is critical: well productivity is dependent on proximity of the well to fracture-correlated lineaments or faults (Allen and Michel, 1998; Mabee, 1999). Boreholes should be fracture-oriented, with fracture locations confirmed by trenching before drilling and by down-hole photography and caliper logs after drilling. Drilling has been found to dilute contaminants in fractures, while pore water is relatively unaffected (McKay et al., 1998). Because of this, some researchers suggest that the extent of contamination is better measured by pore water rather than by fracture water concentrations (McKay et al., 1998). Drilling has also caused cross-contamination, and some researchers use dry-augering to minimize this problem (Nativ et al., 1999).

Improved aquifer-test analytical methods are needed for fractured-rock systems. In quantifying the heterogeneity of a fractured system with extremely variable well yields, a test of over 4,000 wells showed that lithology accounted for 12 percent of the variation. Well construction and aquifer-test duration accounted for 24 percent, with 64 percent of the variation presumably due to (fracture) permeability variation (Knopman and Hollyday, 1993). This test underscores the need for testing and developing field protocols for fractured aquifers.

Methods are needed to locate transmissive fractures. Fractured zones that intersect boreholes can be detected by acoustic televiewers, direct-current resistivity, nuclear magnetic resonance, seismic tomography, electromagnetics, heat-pulse flowmeters, tracers (borehole-to-borehole or single borehole), straddle-packer pump tests, and borehole radar. Heat-pulse flowmeters and thermometers can show which fractures are actively transmitting water even at very low velocities. The acoustic televiewer, which works in muddy water, discriminates bedding-plane and high-angle fractures and reveals conjugate joint sets (Morin et al., 1997). Surface geophysical methods including airborne electromagnetism or infrared airborne spectroscopy may be useful for mapping fracture trends or large individual fractures. Large fractures or fracture zones can be detected by analysis of linear features in air photos, satellite images, and digital elevation model data, using filtering to enhance lineaments. These methods need to be refined and tested, including development both of technology and protocol.

In determination of flow paths, tracers are a promising method for determining fracture-flow. Isotopes, dyes, dissolved chemicals, bacteria, and even lanthanide-labeled clay have been used successfully. Matrix porosity and fracture aperture can be determined with accuracy and are relatively insensitive to type of tracer experiment (Himmelsbach et al., 1998). There is a need for tracer-test protocol. For example, induced-gradient tracer tests may underestimate the importance of dispersion relative to advection because under low-velocity and long-residence-time natural conditions, dispersion dominates transport (Raven et al., 1988). There also needs to be a testing and cataloging of suitable tracers, including natural or isotopic tracers. Parameter-estimation models of fractured systems will help direct data collection; these models have shown that permeability is a poor estimator of fracture aperture, but that flow velocities and tracer breakthrough times are good estimators of aperture (Tsang et al., 1988).

CHARACTERIZATION OF SUBSURFACE HETEROGENEITY

Aquifer heterogeneity arises from the complex history of geologic deposition, erosion, lithification, and tectonic deformation of rocks. The importance of heterogeneity to groundwater occurrence and movement is apparent in the wide range of hydraulic conductivities commonly observed— from 10^{-11} to 10^2 cm/sec (Freeze and Cherry, 1979). Given this range, the determining characteristic of an aquifer in controlling fluid movement is its hydraulic conductivity distribution, or heterogeneity. Despite its importance, characterizing heterogeneity remains elusive.

Scientific and Management Issues

The need for better characterization of heterogeneous aquifers is driven by scientific and public needs for groundwater protection and remediation. Classic hydrogeology has often described aquifers only in terms of bulk hydraulic characteristics (transmissivity, storage coefficient, and porosity) that are relevant to groundwater resources issues. The RASA models, which combined many complex stratigraphic units into a few conceptual layers, are examples of this approach. However, bulk properties are rarely, if ever, adequate to determine flow paths and travel times necessary for contaminant transport studies or wellhead protection. Instead, a more detailed knowledge of the distribution of hydraulic properties is critical.

Efforts to cope with heterogeneity fall into three categories. First, there have been attempts to map heterogeneity by intensive drilling and geophysical surveying. Second, some researchers have attempted to logically relate rock or soil properties to the depositional process, using geologic facies architecture. Facies models—conceptual models of the expected distribution of facies based on the geologic depositional history of an area—can be used to define hydrostratigraphic units (Maxey, 1964; Seaber, 1988; Anderson, 1989). The petroleum industry interprets relatively scarce borehole data and abundant "soft" data such as three-dimensional seismograms using facies models. Third, heterogeneity has been treated as a stochastic process, initially as a purely random distribution of properties, more recently adding realism with correlation, nonstationarity, and nonrandomness.

Predicted hydraulic conductivities look increasingly plausible with these advanced methods, but they still need to be conditioned with information including "soft" data (electrical resistance tomography, seismic tomography, radar tomography, etc.). An abundance of small-scale data are required to detect the underlying stochastic processes for a variety of geologic settings. Detailed studies are needed at heterogeneous sites such as those at the MADE (MAcroDispersion Experiment) site in Mississippi. Stochastic process models will have to be incorporated into facies models to cope with the nonstationarity that appears at the large scale. In the past, detailed characterization usually was not attempted because numerical models, the fundamental tool of modern hydrogeologic prediction, were largely unable to handle this complexity. This situation has changed with the advent of fast, inexpensive computers and improved modeling codes.

Many hydrogeologists have encountered the so-called "scale effect" of hydraulic conductivity (K), which suggests that the effective K of a given material varies with the scale of either the testing method used or the field problem being addressed (Hsieh, 1998). For example, a small-scale contamination study might collect field data and interpret heterogeneity based on wells located only a few meters or tens of meters apart. For a subregional groundwater model (for example, for a small town), heterogeneity might be studied on the scale of hundreds of meters. What is the effective K in these two cases? The question pertains to both the method of measuring K (aquifer test vs. slug test, for example) and the appropriate K to use when simulating aquifer behavior with a numerical model. The number of papers published on the topic of scale since the early 1990s shows there is growing interest in this topic. Different investigators have examined possible causes of the scale effect in several ways, including field-testing and modeling studies. However, there is no consensus on the causes of the effect or on factors that might control its magnitude. Indeed, some hydrogeologists claim there is no physical basis for the scale effect (Butler et al., 1996).

Because heterogeneity results from small-scale (and larger) processes, understanding these processes requires a microscale investigation. Paradoxically, the results will eventually be applied at a larger scale, especially in numerical modeling. So in addition to needing methods to define small-scale features, methods are needed to realistically represent these processes at larger scales.

USGS Roles in Characterization of Subsurface Heterogeneity

The USGS should continue studies of groundwater in a variety of complex settings to reveal important principles and processes controlling water supply and quality. The Survey should also continue its inventory of aquifer properties in order to develop regional databases. The science is by no means complete, as is evident from new developments in the understanding of natural attenuation of contaminants. Translating lithostratigraphy to hydrostratigraphy rests on a foundation of detailed hydrogeologic studies at representative sites such as the Cape Cod toxic waste research site. Detailed studies at sites representative of important (common or especially susceptible to damage) hydrogeologic settings should continue and should be encouraged. It is important, however, that the significance of these studies for generalizing the results to broader areas be understood and emphasized by the USGS and stressed in its reports to the public. The USGS must justify the investment of resources at these intensive-study sites.

In terms of regional groundwater investigations, there is a need for better integration among geologic disciplines: hydrogeology, stratigraphy, sedimentology, and structural geology. The USGS should continue to develop methods of deducing hydrologic information from geologic models and geophysical methods (Jorgensen, 1988). The current Middle Rio Grande basin projects illustrate how this integration can be done successfully. Currently, however, there are no generally accepted "rules" or measures of heterogeneity and its importance; developing such measures would be a fruitful area for research. There are many possible research directions for the improved simulation of the spatial heterogeneity of aquifers (geostatistical models, fractal methods, and process models).

Other areas for investigation include better integration of subsurface stratigraphy with hydrogeology, innovative geophysical tools (downhole logging, geotomography, flowmeters, radar, etc.), measurements of hydraulic parameters such as hydraulic conductivity at a variety of scales, and correlation of these measurements with stratigraphic facies. Tracer experiments, especially experiments that test/verify fieldwork and modeling experiments in heterogeneous aquifers, are needed. It should be noted that the Cape Cod and Borden tests, which have become literature classics, were both conducted at relatively uniform sites.

NUMERICAL MODELING

Scientific and Management Issues

During the last two decades, numerical modeling has become standard practice in most groundwater studies. Better modeling codes, faster and cheaper computers, and user-friendly interfaces have put sophisticated modeling within the facilities and budgets of most groundwater projects (Figure 4.2). However, these advances are a mixed blessing. A 1983 editorial titled "Groundwater Modeling: The Emperor Has No Clothes" (Anderson, 1983) examined the pitfalls of using sophisticated groundwater-flow models without a clear understanding of the modeling process and/or without proper data and model calibration. A follow-up abstract titled "Modeling Complexity: Does the Emperor Have Too Many Clothes?" (Anderson and Hunt, 1998) discussed what has happened to groundwater modeling in the intervening 15 years. The proliferation of model add-ons such as pre- and postprocessors and various optional packages (transport, streamflow routing, lake interactions, evapotranspiration, etc.) has made extremely complex models comparatively easy to construct. Such complex models offer a false sense of ac-

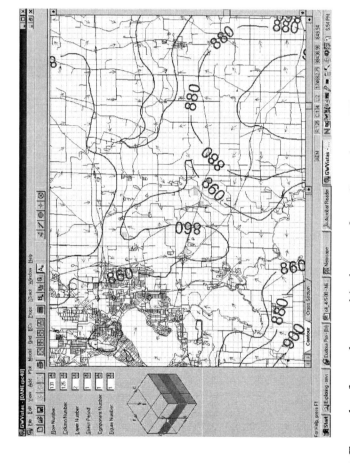

FIGURE 4.2 Example of complex graphical output from Groundwater Vistas, a commercially available graphical user interface for MODFLOW.

curacy and precision if the model complexity cannot be supported with appropriate field information and the model uncertainty is not quantified.

The ability to evaluate uncertainty and sensitivity—an important recent trend in model development—addresses concerns about misleading model results. Parameter estimation codes such as UCODE and MODFLOWP (Poeter and Hill, 1998) allow modelers to estimate optimum sets of model parameters, consistent with field data, and they also provide rigorous measures of the sensitivity of the model solution to changes in parameters. Such uncertainty analyses improve models as tools for decision-making.

Aquifer management-optimization codes—e.g., AQMAN (Lefkoff and Gorelick, 1986), AQMAN3D (Puig et al., 1990), and various commercial products—are a significant step forward in decision-making. Such codes provide optimal groundwater-management solutions, such as the most favorable well placement or pumping rates, under various physical and economic scenarios. They enable, for example, a municipality to maximize groundwater extraction subject to the limitation that heads near a sensitive stream or hazardous waste site do not fall below a threshold value.

Traditional methods of measuring and modeling flow in porous media are being used only cautiously in fractured-rock systems. Significant advances have occurred in the understanding of fractured-rock hydrogeology (NRC, 1996). Most water movement occurs through open fractures, while most storage occurs in the porous matrix. A number of analytical models (e.g., Moench, 1995) now exist for such dual-porosity systems, while sophisticated numerical codes such as FracMan/MAFIC (Golder Associates, 1987) allow evaluation and simulation of discrete fracture networks using stochastic techniques. Field methods are also being developed to characterize these aquifers. One of the major impediments to progress in fractured-rock hydrogeology is a lack of well-characterized field sites for model evaluation. The USGS fractured-rock hydrology research site at Mirror Lake, New Hampshire (Shapiro and Hsieh, 1996), is one of only a few such sites in the United States.

USGS Roles in Numerical Modeling

The USGS has a strong history of innovation and achievement in the development of fundamental groundwater models such as MODFLOW. Efforts should continue in conceptual and theoretical aspects of numerical modeling (flow, reactive chemical transport, management), especially in the near-surface environment, so that increasingly sophisticated models will be available to help diagnose cause-and-effect relationships and perform predictive simulations. However, the committee strongly feels that, in the context of flow modeling, the USGS should devote its efforts to conceptual and theoretical breakthroughs rather than fine-tuning or developing graphical interfaces for codes like MODFLOW. Such work is already being done by the private sector (e.g., Visual MODFLOW, Groundwater Vistas).

In the context of regional groundwater investigations, the USGS should continue to develop appropriate conceptual and numerical "framework" models covering large geographic areas, and it should develop the means for focusing or telescoping these models to smaller scales. The recent work on telescopic mesh refinement (Leake and Claar, 1999) provides examples of such techniques. In addition, analytical element (AE) models can be used for scaling from regional to local simulation. A far-field AE model can be used to develop boundary conditions for a local finite-difference model. Analytical element models have the added advantage of allowing exploration of a model's sensitivity to boundary conditions, an important step that is rarely done (Hunt et al., 1998).

FACILITATING USE OF GROUNDWATER INFORMATION IN DECISION-MAKING

Investigation of these regional issues must provide useful information to water resources managers and decision- or policy-makers. This section discusses three ways that the USGS can assist in this process: (1) by promoting the use of information from USGS studies in decision-making by quantifying and reducing uncertainty in predictions, (2) by scaling results of local studies to the regional level, and (3) by assisting in the development of decision-making and risk models that incorporate groundwater information. The WRD's mission statement clearly emphasizes the need to actively disseminate hydrogeologic data and reports to the public:

The mission of USGS Water Resources Division (WRD) is "to provide reliable, impartial, timely information that is needed to understand the nation's water resources. WRD actively promotes the use of this information by decision-makers to

- Minimize the loss of life and property as a result of water-related hazards, such as floods, droughts, and land movement.
- Effectively manage groundwater and surface-water resources for domestic, agricultural, commercial, industrial, recreational, and ecological uses.
- Protect and enhance water resources for human health, aquatic health, and environmental quality.
- Contribute to the wise physical and economic development of the nation's resources for the benefit of present and future generations." (USGS, 1999c).

Quantifying and Reducing Uncertainty in Predictions

Predictions about groundwater systems are always subject to uncertainty as a result of spatial and temporal variability in subsurface properties and processes. Additional uncertainty arises from attempts to characterize the subsurface based on limited and possibly imprecise measurements. Although uncertainty is an integral part of groundwater systems, past models, measurements, and predictions have not always explicitly identified the associated error.

Future groundwater predictions should specifically include an associated quantitative error. One benefit of estimating error is an improvement in decision-making. Error estimates allow decision-makers and others to understand that hydrologic variables can take on a range of values, facilitating the development of options that will meet objectives under various scenarios. Thus, reporting errors in hydrologic variables should lead to more robust decisions.

Associating uncertainties with predictions and measurements also provides a rational basis for future data collection efforts. Understanding uncertainty and its source allows development of sampling plans that will result in the greatest reductions in uncertainty subject to fiscal and other constraints.

Parameter-estimation modeling provides a measure of uncertainty in predictions that is badly needed. Parameter-estimation modeling should become standard practice, especially when models are used as a basis for water resources decisions. For example, what is the probability that monitoring will detect contamination? If the uncertainty in model results is unacceptable, as it may well be, strategies are needed to diminish that uncertainty.

Scaling Available Information to the Regional Level

How can information that has already been collected at a variety of scales be used in regional-scale studies? Data from past studies are likely to be available on many different scales in new regions of interest to the USGS. Data from smaller-scale or local groundwater studies are likely to have been collected in the past by the Survey and others, and some regional-scale information may be available as well. For example, saturated hydraulic conductivity data may be available from permeameter tests on sediment samples, slug tests, and aquifer tests.

Regional studies will require data collection on regional scales, since many hydrogeologic variables depend upon the measurement scale. However, it makes sense for a regional study to incorporate smaller-scale data previously collected within the region. In regions or parts of regions where hydrogeologic variables are statistically stationary, small-scale parameter values may be representative of larger-scale effective values (Neuzil, 1994). For example, researchers have observed similar values for effective flow parameters on multiple scales at the Mirror Lake site in New Hampshire (Hsieh, 1998). However, as scale changes, new geologic features (fractures, stratigraphic changes) may become important, resulting in regional effective properties that differ from those observed in smaller-scale studies. At some sites, very large changes in permeability have been seen with observation scale (e.g., Bredehoeft et al., 1983).

More research is needed to determine if there are situations in which upscaling (i.e., using data collected on smaller scales to derive information at larger scales) is possible and to develop upscaling methods. Many studies have collected hydrogeologic data on multiple scales; however, researchers may not have taken the further step of developing relationships between the scales. General methods for upscaling have not been established. Indeed, researchers will probably need different methods of upscaling depending on the region's characteristics (homogeneous, stationary, trend/pattern, etc.). Upscaling parameter estimates may not be possible at sites with markedly nonstationary parameter fields unless an observable trend exists (e.g., a linear decrease in permeability with depth).

When a pattern of variability is observed at a number of small-scale studies, researchers sometimes assume that pattern for the larger study. If a number of subregional studies have been conducted in a region, the small-scale studies have clear value as indicators of subregional variability. This information may be particularly important for regional transport studies, where large-scale dispersion is dependent upon small-scale permeability variation.

We recommend that the USGS incorporate into its regional modeling efforts relevant and reliable data collected during previous studies within the regions. Using data from previous studies is particularly important in the groundwater field because of the spatial and temporal variability inherent in subsurface data sets. Because subsurface properties and processes vary in space and time, it may be useful to characterize modeled variables as random or stochastic. Given the impossibility of collecting data everywhere at all times, the properties and processes of interest are always uncertain. In this context, every additional piece of information is valuable in reducing uncertainty in modeling efforts. High data collection costs for the subsurface further increase the value of data available from past studies.

Developing Decision-Making and Risk Models for Groundwater Use

As noted earlier, the WRD should be involved not only in collecting data on water supply, but also in facilitating the use of this information by decision-makers, who have to contend with competing uses (domestic, agricultural, commercial, industrial, recreational, and ecological). Water-use allocation takes into consideration not only scientific knowledge about water resources, but also public policy options, and can be accomplished with the help of models that integrate the two areas (NRC, 1991a). The WRD may, by working in partnership with other national or regional agencies, have a role in analyzing how various policies or laws affect water use regionally. Models could, for example, explore system behavior in response to changes in management or policy, where variables might include cost of pumping water, cost of crop production, income from crops, tax revenues, etc.

CONCLUSIONS

Numerous important advances in hydrogeology have occurred in the past two decades, but serious challenges remain. As part of the Ground-Water Resources Program (GWRP) and associated programs, the USGS WRD should investigate groundwater occurrence and movement in complex hydrogeologic environments such as fractured rock and karst and in heterogeneous media. Advances in theory should be supported by the creative application of field methods and should lead to more realistic models backed by sensitivity and uncertainty analysis.

Surficial aquifers and their boundaries should also receive considerable attention, even in regional studies. Aside from being vulnerable to contamination, shallow aquifers are the focus of research on the spatial and temporal distribution of recharge and discharge and on interactions of groundwater and ecosystems. Many scientific disciplines, including ecology, limnology, chemistry, hydrology, and meteorology, have something to contribute to such groundwater investigations. Regional groundwater studies thus provide ideal opportunities for collaboration within WRD programs and with other USGS divisions and external organizations. Collaboration should also facilitate the development of water-management models, which incorporate legal, economic, ecological, and other constraints.

Finally, changing technology is creating opportunities for innovative approaches to the dissemination of the groundwater information and results generated by such projects. Chapter 5 is devoted to these data issues.

5
Delivery and Accessibility of Groundwater Data

Nationally, the U.S. Geological Survey (USGS) has been and continues to be an important source for credible groundwater data. Regional investigations may involve either the collection of new data and/or the compilation of existing data from state geological surveys, city and county agencies, regional authorities, well drillers, and previous USGS studies. Once assembled, these data have value far beyond their immediate use for a specific study. The opportunity exists to make these compiled data—along with maps and reports generated by the project itself—available to other public and private-sector data users working on local, regional, and national projects. *The posting of physical and chemical groundwater information in easily accessible formats should, therefore, be an integral part of regional studies, despite budgetary pressures.*

In this discussion we distinguish between *raw* or *primary* data, which is the collection of numeric data from various kinds of measurements, and *interpretive* data. The information content of data depends on the current state of understanding of the behavior of the system being measured and on the skill of the interpreter in extracting understanding from the data.

The USGS publishes both primary and interpretative data. Some of the data the USGS publishes (e.g., water-level measurements in wells or piezometers) are only slightly modified from the original measurements. Groundwater data for individual wells are available as time series of these measurements. Interpretation of these data often requires additional information about the site (local and regional pumping rates, etc.). Many of the data published by the USGS are at least partly interpretive.

A common example would be hydraulic conductivity or transmissivity estimates from aquifer-test data. Because methods of analysis change over time, it is essential that the USGS make raw data (such as drawdown as a function of time for an aquifer test), test procedures, and assumptions accessible and linked to the derived data such as transmissivity.

USERS OF GROUNDWATER DATA

Groundwater data users are a diverse group having many different information needs. The same data may have multiple applications, including applications in science, engineering, planning, education, risk assessment, and many other fields. As shown in Table 3-1, users include water resources managers and agencies, agricultural producers, private industry, researchers, federal agencies, policy makers, planners, the media, educational institutions at all levels, environmental groups, state and local cooperators, and private consultants. Although these different user groups may want access to the same data, they may have different interests and abilities to find and manipulate data.

The scale of interest also varies from user to user. A community activist may be interested in the water chemistry from a single industrial site or a suite of community wells, whereas a policy-maker may require information about groundwater resources on regional or even national scales.

The requirements of the users vary vastly as well. Elementary school students studying the hydrologic cycle may make best use of regional or national hydrologic data in an aggregate form as charts, graphs, and tables. A researcher or consultant investigating hazardous chemical migration at a particular location might require primary chemical and hydrologic data with a full array of metadata (e.g., sampling protocol, analytical method, drilling method).

Finally, USGS cooperators often have unique data needs specified in their cooperative agreement with the USGS, such as groundwater-level or chemical data at specific locations. These requirements may or may not be of interest to others, although they are more likely to be broadly useful if compiled into existing databases. Efforts to serve local needs through cooperative agreements should continue to be monitored to ensure that these agreements complement and supplement the national data effort rather than compete with it.

The USGS has the challenge of supplying information to this wide range of users with their diverse data needs. This requires a concerted and broad-based effort to make information delivery—including explanation, demonstration, and follow-through—an integral part of regional studies. Several different media are discussed later in this chapter, but direct forms of communication should not be neglected. Regional projects should include a designated liaison or a liaison committee to meet periodically with the local stakeholders. For example, Survey scientists and their cooperators should strive to participate in public meetings and workshops to explain project results and their implications, to demonstrate data access, and, most importantly, to get feedback on the information needs of the stakeholders.

CONTENT OF GROUNDWATER DATA

Depending on the study, groundwater data may include water-level measurements, water chemistry and water quality parameters, the results of aquifer tests, and other hydrogeologic parameters (e.g., aquifer thickness, mineralogy, recharge rates, porosities, leakage rates, and stream baseflows). Although the content of groundwater data will differ from site to site and from study to study, it should always include good metadata, following accepted metadata standards such as Federal Geographic Data Committee Content Standard for Digital Geospatial Metadata (http://www.fgdc.gov/metadata/contstan.html). Metadata for wells or piezometers should include information such as well location, depth, outer diameter, inner diameter, screened interval, construction details, date drilled, owner's name, and geophysical and lithologic data. Information on local topography, springs, and streams is also important if available. Water-level and water quality measurements should indicate the nature of the measurements and the time the measurements were made so that time series of the data may be generated.

To facilitate data flow between regional and local studies, data should include explicit information on measurement scale. Data from individual observation wells should contain a pointer to available local, aquiferwide, and regional data, including groundwater usage information, hydraulic characteristics, aquifer characteristics, and geologic information. "Effective" values (e.g., the effective transmissivity for an entire aquifer) should include a clear delineation of the area that is included in the analysis, links to the primary data used to develop the effective values, and assumptions made in the data analysis.

FORMAT OF GROUNDWATER DATA

Data-storage technology is rapidly evolving as new and improved means of electronic data storage are made available. A cursory survey of USGS Internet sites shows data available for distribution on paper, microfiche, diskettes, CD-ROM, DVD-ROM, and other media. Some of these media, or their successors, will continue to be used in the future. For example, prepared data sets for educational or outreach purposes are easily prepackaged and mailed on CDs. Paper copies have stood the test of time, are virtually independent of technology changes, and are clearly the best medium for large regional or national maps. Thus, the USGS must continue to provide groundwater data and the products derived from it through a variety of media.

The Internet has revolutionized data delivery. Not only is it fast and efficient, but it also minimizes obsolescence of the transfer medium by transferring the data directly to the user's computer or network. Indeed, as of late 1999, the USGS was serving more than 7 million web pages per month to more than 200,000 users. Of the national surveys in industrialized countries (e.g., British Geological Survey, Geological Survey of Japan, Bureau de Recherches Géologiques et Minières [France], Geological Survey of Canada, and Australian Geological Survey Organization), the USGS appears to be the leader in providing information over the Internet.

Excellent examples of web-based data delivery already exist on the USGS servers. Some of these are products of the USGS Water Resources Division (WRD); others reflect broader efforts within the USGS. These examples are found at both national and regional scales. A number of these web-based data delivery sites are discussed in the next section.

Web-Based Data Sets on a National Scale

Web-based data sets on a national scale include the following:

Delivery and Accessibility of Groundwater Data 103

- *National Atlas of the United States* (http://www.usgs.gov/atlas). The *National Atlas of the United States*—directed by the USGS—is a cooperative effort with many federal and private organizations. Cooperators include the Departments of Interior, Agriculture, and Commerce and the Environmental Systems Research Institute (ESRI). Map layers are grouped by theme or discipline. Under the theme "water" are map layers for dams, watersheds, principal aquifers, real-time streamflow stations, and streams and other water bodies. Under the theme "environment" are layers for Superfund sites, nuclear sites, and many others. Layers can be displayed or downloaded as compressed ArcView shapefiles. Furthermore, the maps are "clickable," so real-time streamflow data for a given site can be accessed directly. Aside from the regional-scale "principal aquifers" layer cited above, however, groundwater information is nonexistent. The availability of groundwater data would improve the "water" theme in the *Atlas*.

- Real-time and historical streamflow data. Real-time and historical streamflow data are provided at two user-friendly sites: http://water.usgs.gov/realtime.html and http://waterdata.usgs.gov/nwis-w/US/. The national maps shown on the main screen in these databases are clickable down to the state or county level, where there is a listing of stream gauges linked directly to downloadable data, metadata, and graphs. These databases are a resource to the nation and represent the type of data presentation we would like to see for groundwater data and other USGS WRD data. At present, there are no links at any level to other kinds of USGS WRD data (water quality data, etc.). Further, no physical or chemical groundwater data are available currently in this format.

- The National Water-Quality Assessment (NAWQA) Program summary of national groundwater and surface water quality data. The NAWQA Program has a summary of national groundwater and surface water quality data on its home page: http://water.usgs.gov/nawqa/. This page currently serves to highlight the individual regional NAWQA studies and the national syntheses that were derived from these studies.

Web-Based Data Sets on Regional Scales

At present, the availability and the quality of regional-scale information vary considerably across the country and are largely dependent on the efforts of individual USGS offices and their state and local cooperators. Most USGS district offices maintain home pages with Internet links to state geological surveys, universities, and regulatory agencies. These USGS district or state web pages can be found indirectly by going to the national web pages and clicking on an individual state or region. Web-based data sets on regional scales include the following:

- Web sites of NAWQA study units. NAWQA study units are by definition regional-scale studies; thus, their web sites present regional data. Some of the NAWQA study units have excellent web sites. The sites can be accessed from clickable maps on the NAWQA home page. The NAWQA regional web sites are of variable quality. Examples of web sites that convey a significant amount of data include the Albemarle–Pamlico NAWQA site (http://sgi1dncrlg.er.usgs.gov/albe-html/ALBEpage.html), which has data, maps, video, pictures, and publications from the study unit, plus educational activities. Other informative sites include the lower Illinois basin NAWQA site (http://www-il.usgs.gov/proj/lirb/). Unlike the surface water sites discussed previously, most maps presently are not georeferenced and are therefore not clickable.
- Regional databases. Excellent regional databases also exist on the web. For example, the USGS Scientific Assessment and Strategy Team (SAST) database on the Upper Mississippi and Lower Missouri river basins (http://edcwww2.cr.usgs.gov/sast-home.html) was designed and built for a study of the flood of 1993 by a team from the USGS, Natural Resources Conservation Service, U.S. Army Corps of Engineers, U.S. Fish and Wildlife Service, U.S. Environmental Protection Agency, and Federal Emergency Management Agency. Although it is now somewhat outdated, this database has different kinds of geological, biological, hydrological, and soil maps that can be viewed online or downloaded in various GIS formats.
- Recent reports of regional studies. Recent reports of regional studies have been published on the web. The USGS Water-Resources Investigations Report 99-4000—*Lithogeochemical Character of Near-Surface Bedrock in the Connecticut, Housatonic, and Thames River Basins* (http://water.usgs.gov/pubs/wri/wri994000/)—demonstrates the potential of the web for directly publishing the results of regional studies. In fact, the web is its principal medium of dissemination. The main product of the report is an ArcInfo-based lithogeochemical (i.e., units

expected to have a characteristic groundwater chemistry) map with associated metadata. The maps can be viewed as PDF files or can be downloaded as ArcInfo export files or ArcView shapefiles. These georeferenced formats would allow the user to overlay layers with other information in the region such as water well chemistry or water levels.

Recommendations for Internet-Based Groundwater Information Delivery

The primary elements for an excellent web-based data and metadata delivery system for groundwater are already in place. Our recommendations would combine the best features of the sites described above.

The *National Atlas of the United States* would make an excellent platform for most groundwater data. The Atlas (1) is run by the USGS but contains a wide variety of information from different agencies, (2) is national in scope, but the layers can also be zoomed down to a state or local scale, (3) is capable of displaying points such as wells, lines such as streams, and polygons such as outlines of regional groundwater project domains or wellhead protection areas, (4) has a thematic structure that would allow related groundwater information to be found and accessed easily, (5) already has a *Real-time Streamflow* layer that could be used as an analogue for well or piezometer data, and (6) is GIS-based so that map layers can not only be seen online, but can also be downloaded for processing with a PC or workstation-based GIS.

Separate map layers would exist delineating ongoing and completed assessments of the NAWQA, Regional Aquifer-System Analysis (RASA), Toxic Substances Hydrology (Toxics), and Federal–State Cooperative Water (Coop) Programs and of regional groundwater programs, with each polygon being linked to a project site containing georeferenced data and maps as well as online reports. Many of these sites already exist and are accessible through the state USGS offices. However, the *Atlas* would simplify the information search for the user by placing the information in the same general location. Older reports might first be available only as scanned images or PDF files, but primary data from these reports should be made available in digital format when feasible. It is also hoped that over time, USGS Water-Resources Investigations Reports could also form a layer in the *Atlas* and be cross-referenced by watershed and county. New projects could be linked from their inception to all of the appropriate layers.

Study results such as maps of modeled water tables or transmissivity distribution should be available not only in formats suitable for viewing online, but also in georeferenced formats that can be read by off-the-shelf GIS packages. This should not be time-intensive, as the major groundwater modeling codes are increasingly able to export results to one or more of these formats.

A layer composed of well locations could be linked to its corresponding water-level data in a way analogous to the existing system for streamflow data. Data on groundwater (and stream) chemistry comprise a major part of the Survey's water resources data collection effort, and these data should be linked with other hydrologic data as technology permits.

National studies could be catalogued in a separate layer in the *Atlas*. The proposed National Aquifer Data Base could also become a part of this system. Thus, information for a given state, county, or watershed at scales ranging from a point to a national summary could be located with minimal difficulty.

In order to facilitate timely information delivery and communication between scientists, most ongoing regional and national investigations with a length of two years or more should establish a project web page that includes a description of the project, identification of project investigators and cooperators, project location, anticipated final and interim products, and data availability. Such web sites should also include, where appropriate, online project reports and links to georeferenced project databases. These project web pages should be directly linked to regional and national data sets through the National Aquifer Data Base. Final reports for all projects should be available digitally in PDF or equivalent portable file formats.

With the easy availability of digital data sets, the USGS and individual scientists working for it will experience increasing pressure by planners, consultants, and others to release data sets, model results, model input files, and other information prior to final publication of the project results and without the lengthy, formal peer-review procedure. It is recommended that the USGS make data available as soon as possible with appropriate disclaimers and metadata documenting the preliminary data released. A precedent for this was established by web publication of the real-time streamflow data.

CONCLUSIONS

The USGS should continue to develop its ability to communicate information regarding regional groundwater systems to decision-makers and the general public through the rapidly developing electronic media as well as through traditional means. Existing USGS web-based templates for surface water data and map layers can be adapted for groundwater data and maps and their metadata. New delivery methods should be publicized on the USGS home page and be backed by technical support from staff of the Earth Science Information Centers. The USGS enjoys a reputation for providing value-neutral free or low-cost primary and interpretive data for public use on natural resources issues. It is essential that this reputation be protected and that the USGS continue to be a reliable and unbiased source of data and information.

6

Conclusions and Recommendations

Groundwater is a vital national resource. Providing accurate and timely information to manage and maintain the quantity and quality of the nation's groundwater requires resource assessments and scientific investigations across a variety of spatial and temporal scales. The U.S. Geological Survey (USGS) is uniquely qualified to conduct and synthesize broad-based investigations. As outlined in an earlier review of the National Water-Quality Assessment (NAWQA) Program (NRC, 1994), the committee believes that regional and national synthesis of information must be a key part of the USGS groundwater program.

The purpose of the regional- and national-scale groundwater investigations envisioned in this report is to provide a framework for decision-makers and a starting point for future site-specific studies by analyzing hydrogeologic processes and conditions relevant to wide areas of the United States. A general characteristic of regional investigations is that they cover hundreds to thousands of square miles and usually extend across state boundaries. However, beyond this commonality, the definition of a "region" varies with the specific problem being addressed. Some regional investigations, such as studies of the High Plains aquifer, may target hydrogeologically distinct, geographically contiguous areas. Other regional studies, such as investigations of salt-water intrusion, may focus on discontinuous but widespread areas, encompassing related nonadjacent aquifer systems sharing common processes.

The committee has reviewed, and is in general agreement with, the *Strategic Directions for the U.S. Geological Survey Ground-Water Resources Program* (USGS, 1998). This document proposes **scientific**

Conclusions and Recommendations 109

assessment of critical groundwater issues combined with a program of **regional and national overviews, access to groundwater data**, and **research and methods development** as key components for heightened work on regional groundwater assessments. The following conclusions and recommendations are intended to enhance this plan and address each of the proposed priorities.

SCIENTIFIC ASSESSMENT OF CRITICAL GROUNDWATER ISSUES

Conclusion: Policy-relevant regional and national assessments of the sustainability of groundwater supplies under current and projected patterns of groundwater use are essential for long-term resource-management decisions. Assessments of sustainability represent the synthesis of resource inventory and characterization, process studies, problem identification, and decision support. The recently published circular *Sustainability of Ground Water Resources* (Alley et al., 1999) shows the wide range of these interlocking issues. The broad topic of sustainability includes the interaction of management decisions (e.g., pumping rates, conjunctive use of groundwater and surface water), resource dynamics (e.g., climatic change, recharge rates), environmental impacts (e.g., streamflow depletion, water quality degradation), and emerging technologies (e.g., aquifer storage and recovery projects).

> **Recommendation:** Regional and national groundwater assessments should have relevance to groundwater sustainability. The management and policy questions that drive regional assessments of sustainability should, in turn, identify and drive the need for regional scientific investigations in fundamental process-oriented groundwater science common to regions with similar problems, settings, or processes. The synthesis of these components on regional and national scales represents a unique activity and should be conducted and coordinated through the groundwater program as a distinct programmatic activity with national relevance.

Conclusion: Within the overarching theme of sustainability, the committee recommends that the following groundwater issues be given the highest priority for investigation by the USGS in the context of regional and national studies. Each issue is relevant over large geographic areas of the United States, has implications for water-management decisions, and raises significant questions best addressed through long-term scientific research.

- *Aquifer management*: optimizing groundwater extraction while limiting undesirable effects such as salt-water intrusion, land subsidence, and harm to ecosystems,
- *Aquifer storage and recovery (ASR) projects*: use of aquifers for repeated storage and recovery of water of varying quality,
- *Groundwater recharge*: quantifying rates, spatial locations, and mechanisms of recharge from local to regional scales,
- *Surficial aquifers*: evaluating hydrogeology, water-level changes, and water quality changes,
- *Interaction of groundwater with surface water*: researching processes and mechanisms in wetlands, rivers, lakes, and coastal areas,
- *Flow and transport in karst and fractured aquifers*: understanding pathways, identifying recharge areas, and characterizing properties at different scales, and
- *Characterization of heterogeneous aquifers at large and small scales*: understanding links between geology and hydrogeology, and developing new characterization methodologies.

Recommendation: The USGS should conduct process studies in these priority issue areas in a regional context as fiscal and human resources permit. These studies should be fully integrated across the Survey's divisions, and they should include aspects of hydrology, geology, biology, and mapping. They must be designed and coordinated to support regional and national synthesis in the context of a national research agenda.

REGIONAL AND NATIONAL OVERVIEWS

Conclusion: A regional and national synthesis effort in the USGS groundwater program must be integrated with other programs to be successful. The current 2 percent of the USGS Water Resources Division (WRD) budget allocated to the groundwater resources program is insuf-

Conclusions and Recommendations

ficient to do much more than identify project areas, coordinate projects, and disseminate information. A meaningful national groundwater program must therefore take advantage of many WRD resources, including the Federal–State Cooperative Water (Coop), National Research, Toxic Substances Hydrology, and NAWQA Programs, and other relevant activities. Studies will necessarily be interdisciplinary. They will likely involve collaborative work with scientists from other divisions of the USGS. Many of the studies will be of such scope and complexity that expertise from outside the USGS may be required. Partnerships with other federal or state science agencies and with research universities through the Water Resources Institutes would likely prove advantageous.

> **Recommendation:** The USGS should coordinate regional investigations across series of related projects of regional significance often having diverse personnel and funding sources. The current Middle Rio Grande basin and Southwest Groundwater projects represent successful models for national synthesis; the USGS should use similar management approaches in other regional-scale investigations.

Conclusion: The USGS needs a clear administrative method for setting priorities and choosing issues and regions for study in the context of national synthesis. The current decision process is not clear to the committee, and it may not be understood by many USGS district staff. Many issues and regions of the United States are appropriate for study, but in an era of limited resources, the USGS needs to make difficult choices in allocating resources to some issues and regions over others. The framework for making these decisions is not addressed in the current WRD *Strategic Plan* (USGS, 1999c). In addition, there is a need for improved coordination of research efforts and data collection with federal, state, local, and private organizations outside the USGS and with universities and state geological surveys as well.

> **Recommendation:** The USGS should implement a formal process for selecting regional- and national-scale groundwater issues for study and for coordinating groundwater research and data collection with other agencies. One way this might be done is through the engagement of advisory groups operating at the state and national levels, and the Survey should allocate funding for the travel, per diem, and other expenses required to operate these committees. Chapter 3 of this report outlines the types of advice and collaboration the committee feels would be appropriate for helping achieve national synthesis. The Survey has taken a similar approach with the Mapping Advisory Council, which aids decision-making for the USGS Mapping Division. At the state level, an appropriate member of the WRD district staff (usually the groundwater section chief) should meet periodically with representatives of state and local regulatory agencies, universities, state geological surveys, and other cooperators in order to determine local priority groundwater issues. The district representative would communicate the conclusions of these meetings back to the national level. Nationally, the USGS should consider forming an advisory committee for regional groundwater investigations. The membership of such a committee might include USGS staff as well as outside members from other federal agencies (U.S. Environmental Protection Agency, Natural Resources Conservation Service, etc.) and from the states (perhaps from the American Association for State Geologists). This committee would meet periodically to assess the findings on the state level and to recommend specific areas and regions for new regional projects.

ACCESS TO GROUNDWATER INFORMATION

Conclusion: Regional groundwater investigations have the complementary goals of developing new scientific knowledge and translating that knowledge into information relevant for decision-making and policy formulation at regional and national levels. The proposed National Aquifer Database, with availability over the Internet, is one mechanism for making study results widely available. However, this database alone will not fill the need for synthesis of the scientific findings; instead, it will provide support and documentation for those findings. The synthesis and the interpretation of findings need to occur on two levels: (1) technical scientific summaries of research findings and (2) interpretive summary and synthesis publications aimed at providing information to decision-makers.

> **Recommendation:** The USGS should continue to produce summary publications and fact sheets appropriate to both technical audiences and decision-makers.

> **Recommendation:** The WRD should hold annual technical meetings or workshops, where various project scientists report their results and project leaders report on future plans and seek input. Such meetings should be organized on local, regional, and national scales.

Conclusion: As building blocks for national synthesis, most of the priority issues outlined in this report will require many studies over many years. Consequently, data management is a critical part of the regionalization effort, especially in an era of rapidly changing data-management and storage technologies.

> **Recommendation:** Digital archiving of modeling and interpretive studies should be a priority in building the cumulative information base upon which regional and national syntheses will be constructed. Archiving must include data, models, codes, documentation, and field techniques.

Conclusion: Now, and in the future, the USGS will be expected to maintain and make available both primary and interpretive groundwater data in digital formats. With the existing USGS Internet sites and such electronic platforms as the *National Atlas of the United States*, the main elements for a web-based data and metadata delivery system are already in place.

> **Recommendation:** In the context of regional and national groundwater studies, the USGS should continue to expand the horizons of electronic delivery of groundwater data in georeferenced formats. The National Aquifer Data Base should be well integrated with the *National Atlas of the United States*, which would make an excellent platform for most groundwater data.

Conclusion: There is an ongoing need for expert interpretation and explanation of groundwater information at the local level. The integration of place-based groundwater studies into regional investigations must ultimately produce information and implications relevant to local water managers and individual citizens. These interpretations are often best provided by individual scientists familiar with the local issues and with local hydrogeology.

> **Recommendation:** The USGS should provide interpretive services at the local and regional levels. Individual scientists at the district level should be encouraged to participate in outreach activities, and they should be rewarded for this participation. USGS management should also ensure that these scientists are aware of other local mechanisms for information dissemination, such as Cooperative Extension offices, soil and water conservation offices, state geological surveys, museums, and local environmental education programs. Continued development of the ability to communicate information about groundwater systems to decision-makers and the general public is important. The USGS must continue to be viewed as a reliable and unbiased source of information.

METHODS DEVELOPMENT

Conclusion: Physically based process models are essential for regionalization of groundwater information. Understanding physical processes, the variations in parameters controlling those processes, and the uncertainty of predictions is essential for investigating groundwater systems at regional scales. The USGS expertise in model development is well known. Recent advances in model visualization, links to GIS, parameter estimation, uncertainty modeling, and optimization provide powerful new tools to the hydrogeologist.

> **Recommendation:** The USGS should continue to explore and develop new modeling approaches—in particular, approaches that improve the understanding of hydrogeologic processes at regional scales. Specific modeling techniques targeted at the priority research areas listed earlier (e.g., fractured rocks, karst, groundwater-surface water interaction, and subsidence) are needed.

Conclusions and Recommendations *115*

Conclusion: One of the most fruitful methods of regionalization and synthesis may be increased emphasis on the links between geology and hydrogeology (e.g., Leahy and Lyttle, 1998). Geology exerts a fundamental control on hydrogeologic processes, and regional variations in geologic setting offer an obvious starting point for regional groundwater studies.

Recommendation: Groundwater investigations at regional scales should generally involve a significant geologic component, with geologists and hydrologists working together on individual projects. The geologic expertise might logically come from colleagues in the USGS Geologic Division as well as from cooperative work with state geological surveys or university personnel.

References

Allen, D. M., and F. A. Michel. 1998. Evaluation of multi-well test data in a faulted aquifer using linear and radial flow models. Ground Water 36:938–948.

Alley, W. M., T. E. Reilly, and O. L. Franke. 1999. Sustainability of Ground-Water Resources. U.S. Geological Survey Circular 1186. Reston, Va.: U.S. Geological Survey.

Amelung, F., D. L. Galloway, J. W. Bell, H. A. Zebker, and R. J. Laczniak. 1999. Sensing the ups and downs of Las Vegas; InSAR reveals structural control of land subsidence and aquifer-system deformation. Geology 27:483–486.

Anderson, M. P. 1983. Ground-water modeling; the emperor has no clothes. Ground Water 21:666–669.

Anderson, M. P. 1989. Hydrogeologic facies models to delineate large-scale spatial trends in glacial and glaciofluvial sediments. GSA Bulletin 101:501–511.

Anderson, M. G., and T. P. Burt. 1980. Interpretation of recession flow. J. Hydrol. 46:89–101.

Anderson, M. P., and R. J. Hunt. 1998. Model complexity: Does the emperor have too many clothes? Eos, Transactions of the American Geophysical Union 79(17):S112.

Bachman, L. J., B. Lindsey, J. Brakebill, and D. S. Powars. 1998. Ground-Water Discharge and Base-Flow Nitrate Loads of Nontidal Streams, and Their Relation to a Hydrogeomorphic Classification of the Chesapeake Bay Watershed, Middle Atlantic Coast. USGS Water-

Resources Investigations Report 98-4059. Reston, Va.: U.S. Geological Survey.

Back, W., and B. B. Hanshaw. 1965. Chemical geohydrology. Pp. 49–109 in Advances in Hydroscience, Vol. 2, V. T. Chow, ed. New York: Academic Press.

Barber, L. B. 1994. Sorption of chlorobenzenes to Cape Cod aquifer sediments. Environmental Science and Technology 28: 890–897.

Barnes, B. S. 1939. The structure of discharge recession curves. Trans. American Geophysical Union 20(4):721–725.

Bartolino, J. R., ed. 1997a. U.S. Geological Survey Middle Rio Grande Basin Study -Proceedings of the First Annual Workshop, Denver, Colorado, November 12–14, 1996. USGS Open-File Report 97-116. Denver, Colo.: U.S. Geological Survey.

Bartolino, J. R. 1997b. Middle Rio Grande Basin Study. USGS Fact Sheet FS-034-97. Reston, Va.: U.S. Geological Survey.

Batiuk, R., P. Heasley, R. Orth, K. Moore, J. C. Stevenson, W. Dennison, L. Staver, V. Carter, N. B. Rybicki, R. E. Hickman, S. Kollar, S. Bieber, and P. Bergstrom. 1992. Chesapeake Bay Submerged Aquatic Vegetation Habitat Requirements and Restoration Goals: A Technical Synthesis. USEPA, Chesapeake Bay Program/Technical Report Series 83/92. Washington, D.C.: U.S. Environmental Protection Agency.

Baydon-Ghyben, W. 1888. Nota in verband met de voorgenomen putboring nabij Amsterdam (Notes on the probable results of well drilling near Amsterdam). Tijdschrift van het Koninklijk Instituut van Ingenieurs, The Hague, 1888/9, 8-22.

Belitz, K. 1999. Impacts of urbanization on groundwater quantity and quality in the Santa Ana Basin, CA. Geol. Soc. of Amer. Abstr. with Prog. 31(7):A-156.

Bencala, K. E., and R. A Walters. 1983. Simulation of solute transport in a mountain pool-and-riffle stream: A transient storage model. Water Resources Res. 19:718–724.

Bögli, A. 1980. Karst Hydrology and Physical Speleology. Translated by J. C. Schmid. Berlin: Springer.

Bredehoeft, J. D., and G. F. Pinder. 1973. Mass transport in flowing groundwater. Water Resources Res. 9:194–210.

Bredehoeft, J. D., C. E. Neuzil, and P. C. D. Milly. 1983. Regional Flow in the Dakota Aquifer – A Study of the Role of Confining Layers. USGS Water-Supply Paper 2237. Reston, Va.: U.S. Geological Survey.

Brinson, M. M. 1993. A Hydromorphic Classification for Wetlands. Wetlands Research Program Technical Report WRP-DE-4. Vicksbury, Miss.: U.S. Army Corps of Engineers.

Butler, J. J., Jr., C. D. McElwee, and W. Liu. 1996. Improving the quality of parameter estimates obtained by slug tests. Ground Water 34: 480–490.

Campana, M. E., and R. M. Boyer, Jr. 1996. A conceptual evaluation of regional ground-water flow, southern Nevada-California, USA. Environmental and Engineering Geoscience II(4):465–478.

Carter, V. G., P. T. Gammon, and N. C. Bartow. 1983. Submersed Aquatic Plants of the Tidal Potomac River. USGS Bull. 1543. Reston, Va.: U.S. Geological Survey.

Carter, V. G., G. Mulamoottil, B. G. Warner, and E. A. McBean. 1996. Environmental gradients, boundaries, and buffers: An overview. Pp. 9–17 in Wetlands: Environmental Gradients, Boundaries, and Buffers, G. Mulamoottil, ed. Boca Raton, Fla.: CRC Press.

Chamberlin, T. C. 1885. The requisite and qualifying conditions of artesian wells. USGS Fifth Annual Report, p. 131–173. Reston, Va.: U.S. Geological Survey.

Chanton, J., J. Baurer, P. Glaser, D. I. Siegel, C. Kelly, S. C. Tyler, E. Romanowicz, and A. Lazarus. 1995. Radiocarbon evidence for the substrates supporting methane formation within northern Minnesota peatlands. Geochem. et Cosmochim. Acta 59:3663–3668.

Chapelle, F. H. 1997. The Hidden Sea. Tucson, Ariz.: Geoscience Press.

Chapelle, F. H. 1999. Bioremediation of petroleum hydrocarbon-contaminated ground water: The perspectives of history and hydrology. Ground Water 37:122–132.

Clark, I., and P. Fritz. 1997. Environmental Isotopes in Hydrogeology. Boca Raton, Fla.: Lewis Publishers.

Cook, P. G., I. D. Jolly, F. W. Leaney, G. R. Walker, G. L. Allan, L. K. Fifield, and G. B. Allison. 1994. Unsaturated zone tritium and chlorine-36 profiles from southern Australia: Their use as tracers of soil water movement. Water Resources Res. 30:1709–1719.

Correll, D. L., T. E. Jordan, and D. E. Weller. 1992. Nutrient flux in a landscape: Effects of coastal land use and terrestrial community mosaic on nutrient transport to coastal waters. Estuaries 15:431–442.

Cowardin, L. M., V. Carter, F. C. Golet, and E. T. LaRoe. 1979. Classification of Wetlands and Deep Water Habitats of the United States. FWS/OBS-79/31. Washington, D.C.: Fish and Wildlife Service.

Dahm, C. N., N. B. Grimm, P. Marmonier, H. M. Valett, and P. Vervier. 1998. Nutrient dynamics at the interface between surface waters and groundwaters. Freshwater Biology 40:427–451.

Daniels, S.H., 2000. Sweet Rewards. The Source 3(1):71-74.

Davis, S. N., and R. J. M. DeWiest. 1966. Hydrogeology. New York: John Wiley and Sons.

Davis, S. N., D. O. Whittemore, and J. Fabryka-Martin. 1998. Uses of chloride/bromide ratios in studies of potable water. Ground Water 36:338–350.

Dennison, W. C., R. J. Orth, K. A. Moore, J. C. Stevenson, V. Carter, S. Kollar, P. W. Bergstrom, and R. A. Batiuk. 1993. Assessing water quality with submersed aquatic vegetation: Habitat requirements as barometers of Chesapeake Bay health. BioScience 43(2):86–94.

Dugan, J. T., and D. A. Cox. 1994. Water-level changes in the High Plains aquifer: Predevelopment to 1993. USGS Water-Resources Investigations Report 94-4157. Reston, Va.: U.S. Geological Survey.

Dugan, J. T., and J. B. Sharpe. 1994. Water-level changes in the High Plains aquifer, 1980 to 1994. USGS Fact Sheet FS-215-95. Reston, Va.: U.S. Geological Survey.

Duncan, D., D. T. Pederson, T. R. Shepherd, and J. D. Carr. 1991. Atrazine used as a tracer of induced recharge. Ground Water Monitoring Review11:144–150.

Essaid, H. I. 1990. A multilayered sharp interface model of coupled freshwater and saltwater flow in coastal systems: Model development and application. Water Resources Res. 26:1431–1454.

Fetter, C. W. 1994. Applied Hydrogeology. 3rd edition. New York: Macmillan.

Focazio, M. J., L. N. Plummer, J. K. Bohlke, E. Busenberg, L. J. Bachman, and D. S. Powars. 1998. Preliminary Estimates of Residence Times and Apparent Ages of Ground Water in the Chesapeake Bay Watershed, and Water-Quality Data from a Survey of Springs. USGS Water-Resources Investigations Report 97-4225. Reston, Va.: U.S. Geological Survey.

Freeze, R. A., and J. A. Cherry. 1979. Groundwater. Englewood Cliffs, N.J.: Prentice Hall.

Freeze, R. A., and P. A. Witherspoon. 1967. Theoretical analysis of regional groundwater flow. II: Effect of water table configuration and subsurface permeability variations. Water Resources Res. 3:623–634.

Garcia, D. H. 1998. Competition between public agencies and the private sector. The Professional Geologist 35(13): 9–10.

Gibert, J., M. J. Dole-Olivier, P. Marmonier, and P. Vervier. 1990. Surface water/groundwater ecotones. Pp. 199–225 in Ecology and Management of Aquatic-Terrestrial Ecotones, R. J. Naiman and H. Decamps, eds. Man and the Biosphere Series, Vol. 4. Paris: UNESCO.

Golder Associates. 1987. FracMan software suite and MAFIC (Matrix and Fracture Interaction Code). Redmond, Wash.: Golder Associates.

Guerrero, P. F. 1999. Superfund: Progress, Problems, and Future Outlook. Testimony on March 23, 1999, before the Finance and Hazardous Materials Subcommittee, House Committee on Commerce (T-RCED-99-128).

Gutentag, E. D., F. J. Heimes, N. C. Krothe, R. R. Luckey, and J. B. Weeks. 1984. Geohydrology of the High Plains Aquifer in Parts of Colorado, Kansas, Nebraska, New Mexico, Oklahoma, South Dakota, Texas, and Wyoming. USGS Professional Paper 1400-B. Reston, Va.: U.S. Geological Survey.

Harr, J. 1995. A Civil Action. New York: Random House.

Harte, P.T., and T. C. Winter. 1995. Simulations of flow in crystalline rock and recharge from overlying glacial deposits in a hypothetical New England setting: Ground Water 33:953–964.

Harvey, R. W. 1993. Fate and transport of bacteria injected into aquifers. Current Opinion in Biotechnology 4:312–317.

Harvey, J. W., and K. E. Bencala. 1993. The effect of streambed topography on surface-subsurface water exchange in mountain catchments. Water Resources Res. 29:89–98.

Harvey, J. W., B. J. Wagner, and K. E. Bencala. 1996. Evaluating the reliability of the stream tracer approach to characterize stream-subsurface water exchange. Water Resources Res. 32:2441–2451.

Hawley, J. W., and C. S. Haase. 1992. Hydrogeologic framework of the northern Albuquerque Basin. New Mexico Bureau of Mines and Mineral Resources Open-File Report 387. Socorro: NMBMMR.

Hem, J. D. 1959. Study and Interpretation of the Chemical Characteristics of Natural Water. USGS Water Supply Paper 1473. Reston, Va.: U.S. Geological Survey.

References

Hendry, M. J., J. R. Lawrence, and P. Maloszewski. 1999. Effects of velocity on the transport of two bacteria through saturated sand. Ground Water 37:103.

Himmelsbach, T., H. Hötzl, and P. Maloszewski. 1998. Solute transport processes in a highly permeable fault zone of Lindau fractured rock test site (Germany). Ground Water 36:792.

Hsieh, Paul A. 1998. Scale effects in fluid flow through fractured geologic media. Pp. 335–353 in Scale Dependence and Scale Invariance in Hydrology, G. Sposito, ed. New York: Cambridge University Press.

Hubbert, J. K. 1940. The theory of groundwater motion. J. Geol. 48:785–944.

Hunt, R. J., M. P. Anderson, and V. A. Kelson. 1998. Improving a complex finite-difference ground water flow model through the use of an analytic element screening model. Ground Water 36:1011–1017.

Hunt, R. J., D. P. Krabenhoft, and M. P. Anderson. 1997. Assessing hydrogeochemical heterogeneity in natural and constructed wetlands. Biogeochemistry 39:271–293.

Izbicki, J. A. 1996. Seawater Intrusion in a Coastal California Aquifer. USGS Fact Sheet 125-96. Reston, Va.: U.S. Geological Survey. Available online at http://ca.water.usgs.gov-fact/b07/ (accessed June, 2000).

Johnston, R. H. 1997. Hydrologic Budgets of Regional Aquifer Systems of the United States for Predevelopment and Development Conditions. USGS Professional Paper 1425. Reston, Va.: U. S. Geological Survey.

Johnston, C. A., N. E. Detenbeck, and G. J. Niemi. 1990. The cumulative effect of wetlands on stream water quality and quantity: A landscape approach. Biogeochemistry 10:105–141.

Jorgensen, D. G. 1988. Using Geophysical Logs to Estimate Porosity, Water Resistivity, and Intrinsic Permeability. USGS Water-Supply Paper 2321. Reston, Va.: U.S. Geological Survey.

Kazmann, R. G. 1972. Modern Hydrology. 2nd ed. New York: Harper and Row.

Kernodle, J. M., D. P. McAda, and C. R. Thorn. 1995. Simulation of ground-water flow in the Albuquerque Basin, central New Mexico, with projections to 2020. U.S. Geological Survey Water-Resources Investigations Report 94-4251. Reston, Va.: U.S. Geological Survey.

King, F. H. 1899. Principles and Conditions of the Movements of Groundwater. USGS 19th Annual Report, Part 2, pp. 59–294. Reston, Va.: U.S. Geological Survey.

Knopman, D. S., and E. F. Hollyday. 1993. Variation in specific capacity in fractured rocks, Pennsylvania. Ground Water 31:135–145.

Konikow, L. F., and J. D. Bredehoeft. 1978. Computer Model of Two-Dimensional Solute Transport and Dispersion in Ground Water. U.S. Geological Survey Techniques of Water-Re-sources Investigations, Book 7, Chapter C2. Reston, Va.: U.S. Geological Survey.

Kreitler, C. W. 1977. Fault control of subsidence, Houston, Texas. Ground Water 15(3):203–214.

Krueger, C. J., L. B. Barber, D. W. Metge, and J. A. Field. 1998. Fate and transport of linear alkylbenzenesulfonate in a sewage-contaminated aquifer: A comparison of natural-gradient pulsed tracer tests. Environmental Science and Technology 32:1134–1142.

Ku, H. F. H., and D. B. Aaronson. 1992. Rates of water movement through the floors of selected stormwater basins in Nassau County, Long Island, New York. USGS Water-Resources Investigations Report 91-4012. Reston, Va.: U.S. Geological Survey.

LaBaugh, J. W., T. C. Winter, D. O. Rosenberry, P. F. Schuster, M. M. Reddy, and G. R. Aiken. 1997. Hydrological and chemical estimates of the water balance of a closed-basin lake in north central Minnesota. Water Resources Res. 33:2799–2812.

LaBaugh, J. W., T. C. Winter, G. A. Swanson, D. O. Rosenberry, R. D. Nelson, and N. H. Euliss, Jr. 1996. Changes in atmosphere circulation patterns affect mid-continent wetlands sensitive to climate. Limnology and Oceanography 41:864–970.

Leahy, P. P., and P. T. Lyttle. 1998. The re-emerging and critical role of geologic understanding in hydrogeology. Pp. 19–24 in the proceedings of the joint meeting of the XXVIII Congress of the International Association of Hydrogeologists and the Annual meeting of the American Institute of Hydrologists, J. V. Brahana et al., eds. St. Paul, MN: American Institute of Hydrology.

Leahy, P. P., and W. G. Wilber. 1991. National Water-Quality Assessment Program. USGS Open-File Report 91-54. Reston, Va.: U.S. Geological Survey.

Leake, S. A., and D. V. Claar. 1999. Computer Programs and Procedures for Telescopic Mesh Refinement Using MODFLOW. USGS Open-File Report 99-238. Reston, Va.: U.S. Geological Survey.

Lee, C. H. 1915. The determination of safe yield of underground reservoirs of the closed basin type. Trans. Am. Soc. Civil Engrs. 78:148–151.
Lefkoff, L. J., and S. M. Gorelick. 1986. AQMAN: Linear and Quadratic Programming Matrix Generator Using Two-Dimensional Ground Water Flow Simulation for Aquifer Management Modelling. USGS Water Resources Investigation 86-4016. Reston, Va.: U.S. Geological Survey.
Mabee, S. B. 1999. Factors influencing well productivity in glaciated metamorphic rocks. Ground Water 37:88-97.
Massonnet, D., and K. L. Feigl. 1998. Radar interferometry and its application to changes in the earth's surface. Reviews of Geophysics 36:441–500.
Maxey, G. B. 1964. Hydrostratigraphic units. J. Hydrol. 2: 124–129.
Mazur, A. 1998. A Hazardous Inquiry: The Rashomon Effect at Love Canal. Cambridge, Mass.: Harvard University Press.
McConnaughey, T. A., F. Doumenge, D. Allemand, and A. Toulemont. 1994. Calcification, photosynthesis, and global carbon cycles. In Past and Present Biomineralization Processes: Considerations About the Carbonate Cycle, F. Doumenge, ed. Bulletin de la Institut Oceanographique (Monaco) 13:137–161.
McDonald, M. G., and A. W. Harbaugh. 1988. A Modular Three-Dimensional Finite-Difference Ground-Water Flow Model. U.S. Geological Survey Techniques of Water-Resources Investigations. Book 6, Chapter A1. Reston, Va.: U.S. Geological Survey.
McKay, L. D., D. J. Balfour, and J. A. Cherry. 1998. Lateral chloride migration from a landfill in a fractured clay-rich glacial deposit. Ground Water 36: 988–999.
Meinzer, O. E. 1923. Outline of Groundwater in Hydrology with Definitions. USGS Water Supply Paper 494. Reston, Va.: U.S. Geological Survey.
Merritt, M. L. 1996. Assessment of Saltwater Intrusion in Southern Coastal Broward County, Florida. USGS Water-Resources Investigations Report 96-4221. Reston, Va.: U.S. Geological Survey.
Mitsch, W. J., and J. G. Gosselink. 1993. Wetlands. 2nd ed. New York: Van Nostrand Reinhold.
Moench, A. F. 1995. Convergent radial dispersion in a double-porosity aquifer with fracture skin: Analytical solution and application to a field experiment in fractured chalk. Water Resources Res. 31:1823–1835.

Molnia, B. F. 1999. Meet Charles Groat, director of the U. S. Geological Survey. GSA Today 9(8):10–13.
Moore, M. H. 1995. Creating Public Value: Strategic Management in Government. Cambridge, Mass.: Harvard University Press.
Morin, R. H., G. B. Carleton, and S. Poirier. 1997. Fractured-aquifer hydrogeology from geophysical logs: The Passaic Formation, New Jersey. Ground Water 35:328–338.
Morrice, J. A., H. M. Valett, C. N. Dahm, and M. E. Campana. 1997. Alluvial characteristics, groundwater-surface water exchange and hydrological retention in headwater streams. Hydrologic Processes 11:253–267.
National Research Council (NRC). 1988. Hazardous Waste Site Management: Water Quality Issues. Washington, D.C.: National Academy Press.
National Research Council (NRC). 1990. Ground Water Models: Scientific and Regulatory Applications. Washington, D.C.: National Academy Press.
National Research Council (NRC). 1991a. Preparing for the Twenty-First Century: A report to the USGS Water Resources Division. Washington, D.C.: National Academy Press.
National Research Council (NRC). 1991b. Opportunities in the Hydrologic Sciences. Washington, D.C.: National Academy Press.
National Research Council (NRC). 1993. Ground Water Vulnerability Assessment: Predicting Relative Contamination Potential Under Conditions of Uncertainty. Washington, D.C.: National Academy Press.
National Research Council (NRC). 1994. National Water Quality Assessment Program: The Challenge of a National Synthesis. Washington, D.C.: National Academy Press.
National Research Council (NRC). 1995a. Mexico City's Water Supply: Improving the Outlook for Sustainability. Washington, D.C.: National Academy Press.
National Research Council (NRC). 1995b. Wetland Characteristics and Boundaries. Washington, D.C.: National Academy Press.
National Research Council (NRC). 1996. Hazardous Materials in the Hydrologic Environment: The Role of Research by the U.S. Geological Survey. Washington, D.C.: National Academy Press.
National Research Council (NRC). 1997a. Building a Foundation for Sound Environmental Decisions. Washington, D.C.: National Academy Press.

National Research Council (NRC). 1997b. Valuing Ground Water: Economic Concepts and Approaches. Washington, D.C.: National Academy Press.

National Research Council (NRC). 1997c. Safe Water from Every Tap: Improving Water Service to Small Communities. Washington, D.C.: National Academy Press.

National Research Council (NRC). 1998. Issues in Potable Reuse: The Viability of Augmenting Drinking Water Supplies with Reclaimed Water. Washington, D.C.: National Academy Press.

Nativ, R., E. M. Adar, and A. Becker. 1999. Designing a monitoring network for contaminated ground water in fractured chalk. Ground Water 37:38–47.

Neuzil, C. E. 1994. How permeable are clays and shales? Water Resources Res. 30:145–150.

Nolan, B. T., B. C. Ruddy, K. J. Hitt, and D. R. Helsel. 1998. A national look at nitrate contamination of ground water. Water Conditioning and Purification 39:76–79.

Novitski, R. P. 1979. Hydrologic characteristics of Wisconsin's wetlands and their influence on floods, stream flow, and sediment. Pp. 377–388 in Wetland Functions and Values: The State of Our Understanding, P. E. Greeson, J. R. Clark, and J. E. Clark, eds. Minneapolis, Minn.: American Water Resources Association.

Parkhurst, D. L., D. C. Thorstenson, and L. N. Plummer. 1980. PHREEQE—A Computer Program for Geochemical Calculations. USGS Water-Resources Investigations Report 80-96. Reston, Va.: U.S. Geological Survey.

Person, M., J. Z. Taylor, and S. L. Dingman. 1998. Sharp interface models of salt water intrusion and wellhead delineation on Nantucket Island, Massachusetts. Ground Water 36:731.

Pinder, G. F., and J. D. Bredehoeft. 1968. Application of the digital computer for aquifer evaluation. Water Resources Res. 4:1069–1093.

Poeter, E. P., and M. C. Hill. 1998. Documentation of UCODE, a Computer Code for Universal Inverse Modeling. USGS Water-Resources Investigations Report 98-4080. Reston, Va.: U.S. Geological Survey.

Poiani, K. A., B. L. Bedford, and M. D. Merrill. 1996. A GIS-based index for relating landscape characteristics to potential nitrogen leaching to wetlands. Landscape Ecology 11(4):237.

Poland, J. F., B. E. Lofgren, R. L. Ireland, and R. G. Pugh. 1975. Land subsidence in the San Joaquin Valley, California, as of 1972. U.S. Geological Survey Professional Paper 437-H. Reston, VA.: U.S. Geological Survey.

Pollock, D. W. 1989. Documentation of Computer Programs to Compute and Display Pathlines Using Results from the U.S. Geological Survey Modular Three-Dimensional Finite-Difference Ground-Water Flow Model. USGS Open File Report 89-131. Reston, Va.: U.S. Geological Survey.

Pope, D. A., and A. D. Gordon. 1999. Simulation of Ground-Water Flow and Movement of the Freshwater-Saltwater Interface in the New Jersey Coastal Plain. USGS Water-Resources Investigations Report 98-4216. Reston, Va.: U.S. Geological Survey.

Puig, J. C., L. I. Rolón-Collazo, and Ismael Pagán-Trinidad. 1990. Development of an aquifer management model: AQMAN3D. Pp. 39–48 in Tropical Hydrology and Caribbean Water Resources, J. H. Krishna, V. Quinones-Aponte, F. Gomez-Gomez, and G. L. Morris, eds. Proceedings of the International Symposium on Tropical Hydrology, San Juan, Puerto Rico, July 23–27, 1990, American Water Resources Association Technical Proceeding Series 90-2.

Pyne, R. D. G. 1995. Groundwater Recharge and Wells: A Guide to Aquifer Storage Recovery. Boca Raton, Fla.: CRC Press - Lewis Publishers.

Raven, K. G., K. S. Novakowski, and P. A. Lapcevic. 1988. Interpretation of field tracer tests of a single fracture using a transient solute storage model. Water Resources Res. 24:2019–2032.

Reichard, E. G. 1995. Groundwater-surface water management with stochastic surface water supplies: A simulation optimization approach. Water Resources Res. 31:2845–2865.

Reisner, M. 1993. Cadillac Desert: The American West and Its Disappearing Water. New York: Penguin.

Rice, K. C., and G. M. Hornberger. 1998. Comparison of hydrochemical tracers to estimate source contributions to peak flow in a small, forested, headwater catchment. Water Resources Res. 34:1755–1766.

Roadcap, G. S., M. B. Wentzel, S. D. Lin, E. E. Herricks, R. K. Raman, R. L. Locke, and D. L. Hullinger. 1999. An assessment of the hydrology and water quality of Indian Ridge marsh and the potential effects of wetland rehabilitation on the diversity of wetland plant communities, Illinois State Water Survey Contract Report 624. Champaign, IL: ISWS.

Robinson, N. I., J. M. Sharp, Jr., and I. Kreisel. 1998. Contaminant transport in sets of parallel finite fractures with fracture skins. J. Contaminant Hydrology 31:83–109.

Romanowicz, E. A., D. I. Siegel, J. P. Chanton, and P. H. Glaser. 1995. Temporal variations of deep dissolved-methane in the Lake Agassiz Peatland. Global Biogeochemical Cycles 9:197–212.

Rosenberry, D. O., and T. C. Winter. 1997. Dynamics of water-table fluctuations in an upland between two prairie-pothole wetlands in North Dakota. J. Hydrol. 191:266–289.

Rosenberry, D. O., J. W. LaBaugh, T. M. McConnaughey, R. G. Striegl, and T. C. Winter. 1993. Hydrologic Research in the Shingobee River Headwaters Area, Minnesota. USGS Open-File Report 93-446. Reston, Va.: U.S. Geological Survey.

Rozycki, A. 1996. Analysis of a marine intrusion by parameters derived from salt-water conductivity. Ground Water 34:1076–1081.

Schwalb, A., S. M. Locke, W. E. Dean, and L. C. K. Shane. 1995. Ostracode ^{18}O and ^{13}C evidence of Holocene environmental changes in the sediments of two Minnesota lakes. Journal of Paleolimnology 14(3):281–296.

Seaber, P. R. 1988. Hydrostratigraphic units. Pp. 9–14 in Hydrogeology, W. Back, J. S. Rosenshein, and P. R. Seaber, eds. Washington, D.C.: Geological Society of America.

Shapiro, A., and P. Hsieh. 1996. Overview of research on use of hydrologic, geophysical, and geochemical methods to characterize flow and chemical transport in fractured rock at the Mirror Lake site, New Hampshire. Pp. 71–80 in USGS Water-Resources Investigations Report 94-4015, Vol. 1, D. W. Morganwalp and D. A. Aronson, eds. Reston, Va.: U.S. Geological Survey.

Shotyk, W., D. Weiss, P. G. Appleby, A. K. Cheburkin, R. Frei, M. Gloor, J. D. Kramers, S. Reese, and W. O. van der Knaap. 1998. History of atmospheric lead deposits since 12,370 ^{14}C yr BP from a peat bog, Jura Mountains, Switzerland. Science 281:1635–1640.

Siegel, D. I. 1988. The recharge-discharge functions of wetlands near Juneau, Alaska: Part I, Hydrogeological investigations. Ground Water 26:427–434.

Simmons, C. T., K. A. Narayan, and R. A. Wooding. 1999. On a test case for density-dependent groundwater flow and solute transport models: The salt lake problem. Water Resources Res. 35:3607–3620.

Sklash, M. G., and R. N. Farvolden. 1979. The role of groundwater in storm runoff. J. Hydrol. 43:45–65.

Slate, J. L., ed. 1998. U.S. Geological Survey Middle Rio Grande Basin Study. USGS Open-File Report 98-337. Denver, Colo.: U.S. Geological Survey.

Slichter, C. S. 1902. The Motions of Underground Waters. USGS Water-Supply Paper 67. Reston, Va.: U.S. Geological Survey.

Smith, B. S. 1994. Saltwater Movement in the Upper Floridan Aquifer Beneath Port Royal Sound, South Carolina. USGS Water-Supply Paper 2421. Reston, Va.: U.S. Geological Survey.

Smith, R. L., M. L. Ceazan, and M. H. Brooks. 1994. Autotrophic, hydrogen-oxidizing, denitrifying bacteria in groundwater: Potential agents for bioremediation of nitrate contamination. Applied and Environmental Microbiology 60:1949–1955.

Solley, W. B., R. B. Pierce, and H. A. Perlman. 1998. Estimated Use of Water in the United States in 1995. USGS Circular 1200. Reston, Va.: U.S. Geological Survey.

Sophocleous, M. A. 1992. Groundwater recharge estimation and regionalization: The Great Bend Prairie of central Kansas and its recharge statistics. Journal of Hydrology 127:113–140.

Sophocleous, M. A., and T. Ma. 1998. A decision support model to assess vulnerability to salt water intrusion in the Great Bend Prairie aquifer of Kansas. Ground Water 36:476–483.

Squillace, P. J. 1996. Observed and simulated movement of bank-storage water. Ground Water 34:121–134.

Squillace, P. J., E. M. Thurman, and E. T. Furlong. 1993. Groundwater as a nonpoint source of atrazine and diethylatrazine in a river during base flow conditions. Water Resources Res. 29:1719–1729.

Stanford, J., and J. V. Ward. 1993. An ecosystem perspective of alluvial rivers: Connectivity and the hyporheic corridor. J. North Amer. Bentho. Soc. 12:48–60.

Steinmann, P., and W. Shotyk. 1997. Geochemistry, mineralogy, and geochemical mass balance of major elements in two peat bog profiles (Jura Mountains, Switzerland). Chemical Geology 138(1-2):25–53.

Strack, O. D. L. 1989. Groundwater Mechanics. Englewood Cliffs, N.J.: Prentice Hall.

Sun, R. J., and R. H. Johnston. 1994. Regional Aquifer-System Analysis Program of the U.S. Geological Survey, 1978–1992. USGS Circular 1099. Reston, Va.: U.S. Geological Survey.

Sun, R. J., J. B. Weeks, and H. F. Grubb. 1997. Bibliography of Regional Aquifer-System Analysis Program of the U.S. Geological Survey, 1978–96. USGS Water-Resources Investigations Report 97-4074. Reston, Va.: U.S. Geological Survey.

Swarzenski, P. W. 1999. Examining Freshwater–Saltwater Interface Processes with Four Radium Isotopes. USGS Fact Sheet FS-065-99. Reston, Va.: U.S. Geological Survey.

Tarhouni, J., and L. Lebbe. 1996. Optimization of recharge and pumping rates by means of an inverse 3D model. Water Resources Management 10:355–371.

Texas Natural Resource Conservation Commission (TNRCC). 1997. Evaluation of Existing Water Availability Models, Technical Paper No. 2. Austin, TX: TNRCC.

Theis, C. V. 1935. The relation between the lowering of the piezometric surface and rate and duration of discharge of a well using groundwater storage. Trans. Am. Geophys. Union 2:519–524.

Thorn, C. R., D. P. McAda, and J. M. Kernodle. 1993. Geohydrologic framework and conditions in the Albuquerque Basin, Central New Mexico. USGS Water-Resources Investigations Report 93-4149. Reston, Va.: U.S. Geological Survey.

Todd, D. K. 1959. Groundwater Hydrology. 1st ed. New York: John Wiley and Sons.

Tomaszewski, D. J. 1996. Distribution and Movement of Saltwater in Aquifers in the Baton Rouge Area, Louisiana, 1990–1992. Louisiana Dept. of Transportation and Development Water Resources Technical Report 59. Baton Rouge, LA: LDTD.

Tóth, J. 1962. A theory of groundwater motion in small drainage basins in Central Alberta. J. Geophys. Res. 67:4375–4387.

Triska, F. J., V. C. Kennedy, R. J. Avanzino, G. W. Zellweger, and K. E. Bencala. 1989. Retention and transport of nutrients in a third-order stream in northwestern California: Hyporheic processes. Ecology 70:1893–1905.

Truesdell, A. H., and B. F. Jones. 1974. WATEQ: A computer program for calculating chemical equilibria of natural waters. Journal of Research 2:233–274.

Tsang, Y. W., C. F. Tsang, I. Neretnieks, and L. Moreno. 1988. Flow and transport in fractured media: A variable aperture channel model and its properties. Water Resources Res. 24:2049–2060.

U. S. Geological Survey (USGS). 1995. Ground-Water Quality in the Calumet Region of Northwestern Indiana and Northeastern Illinois. USGS Water-Resources Investigations Report 95-4244. Reston, Va.: U.S. Geological Survey.

U. S. Geological Survey (USGS). 1998. Strategic Directions for the U.S. Geological Survey Ground-Water Resources Program. Report to Congress, November 30, 1998. Reston, Va.: U.S. Geological Survey.

U. S. Geological Survey (USGS). 1999a. Land Subsidence in the United States. USGS Circular 1182. Reston, Va.: U.S. Geological Survey.

U. S. Geological Survey (USGS). 1999b. The Quality of Our Nation's Waters: Nutrients and Pesticides. USGS Circular 1225. Reston, Va.: U.S. Geological Survey.

U. S. Geological Survey (USGS). 1999c. Strategic Directions for the Water Resources Division, 1998–2008. USGS Open-File Report 99-249. Reston, Va.: U.S. Geological Survey.

Valett, H. M., C. N. Dahm, M. E. Campana, J. A. Morrice, M. A. Baker, and C. S. Fellows. 1997. Hydrologic influences on groundwater–surface water ecotones: Heterogeneity in nutrient composition and retention. J. North Amer. Bentho. Soc. 16:239–247.

Valett, H. M., J. A. Morrice, C. N. Dahm, and M. E. Campana. 1996. Parent lithology, surface-groundwater exchange and nitrate retention in headwater streams. Limnology and Oceanography 41:333–345.

Valiela I., K. Foreman, M. LaMontagne, D. Hersh, J. Costa, P. Peckol, B. DeMeo-Anderson, C. D'Avanzo, M. Babione, C. H. Sham, J. Brawley, and K. Lajtha. 1992. Couplings of watersheds and coastal waters: Sources and consequences of nutrient enrichment in Waquoit Bay, Massachusetts. Estuaries 15:433–457.

Vengosh, A., and I. Pankratov. 1998. Chloride/bromide and chloride/fluoride ratios of domestic sewage effluents and associated contaminated ground water. Ground Water 36:815–825.

Verhoest, N. E. C., F. P. De Troch, and P. A. Troch. 1998. Analysis of soil moisture in agricultural fields based on RADARSAT SLC data. Pp. 13–15 in Proceedings of Radarsat/Application Development and Research Opportunity, Montreal, Canada, October 13-15, 1998. Richmond, British Columbia, Canada: RADARSAT.

Vervier, P., J. Gibert, P. Marmonier, and M. J. Dole-Olivier. 1992. A perspective on the permeability of the surface freshwater–groundwater ecotone. J. North Amer. Bentho. Soc. 11:93–102.

Voss, C. I. 1984. A Finite-Element Simulation Model for Saturated-Unsaturated, Fluid-Density-Dependent Ground Water Flow with Energy Transport or Chemically Reactive Single-Species Solute Transport. USGS Water-Resources Investigations Report 84-4369. Reston, Va.: U.S. Geological Survey.

White, W. A., T. A. Tremblay, E. G. Wermund, Jr., and L. R. Handley. 1993. Trends and status of wetland and aquatic habitats in the Galveston Bay system, Texas. Galveston Bay National Estuary Program Publication GBNEP-31. Webster, TX: GBNEP.

Whiting, P. J., and M. Pomeranets. 1997. A numerical study of bank storage and its contribution to streamflow. J. Hydrol. 202:121–136.

Winter, T. C. 1976. Numerical Simulation Analysis of the Interaction of Lakes and Ground Water. USGS Professional Paper 1001. Reston, Va.: U.S. Geological Survey.

Winter, T. C. 1978. Numerical simulation of steady-state three-dimensional groundwater flow near lakes. Water Resources Res. 14:245–254.

Winter, T. C. 1981. Effects of water-table configuration on seepage through lakebeds. Limnology and Oceanography 26:925–943.

Winter, T. C. 1988. A conceptual framework for assessing cumulative impacts on hydrology of nontidal wetlands. Environmental Management 13:605–620.

Winter, T. C. 1995a. Hydrologic processes and the water budget of lakes. Pp. 37–62 in Physics and Chemistry of Lakes, D. Imboden, J. R. Gat, and A. Lerman, eds. Heidelberg: Springer-Verlag.

Winter, T. C. 1995b. Recent advances in understanding the interaction of groundwater and surface water: Reviews of Geophysics 33 suppl.: 985-994.

Winter, T. C., ed. 1997. Hydrological and Biogeochemical Research in Shingobee River Headwaters Area: North-Central Minnesota. USGS Water-Resources Investigations Report 96-4215. Reston, Va.: U.S. Geological Survey.

Winter, T. C., and M. K. Woo. 1990. Hydrology of lakes and wetlands. Pp. 159–197 in Surface Water Hydrology: The Geology of North America. Boulder, Colo.: Geological Society of America.

Winter, T. C., J. W. Harvey, O. L. Franke, and W. M. Alley. 1998. Ground Water and Surface Water: A Single Resource. USGS Circular 1139. Reston, Va.: U.S. Geological Survey.

Winter, T. C., D. O. Rosenberry, and A. M. Sturrock. 1995. Evaluation of eleven equations for determining evaporation for a small lake in the north-central United States. Water Resources Res. 31: 983–993.

Woessner, W. W. 1998. Changing views of stream-groundwater interaction. Pp. 1–6 in Gambling with Groundwater, J. Brahana et al., eds. St. Paul, Minn.: American Institute of Hydrology.

Wondzell, S. M., and F. J. Swanson. 1996. Seasonal and storm flow dynamics of the hyporheic zone of a 4th-order mountain stream. I. Hydrologic processes. J. North Amer. Bentho. Soc. 15:3–19.

Wroblicky, G. J., M. E. Campana, H. M. Valett, and C. N. Dahm. 1998. Seasonal variation in surface-subsurface water exchange and lateral hyporheic area in two stream-aquifer systems. Water Resources Res. 34:317–328.

Young, R. A., and J. D. Bredehoeft. 1972. Digital computer simulation for solving management problems of conjunctive groundwater and surface water systems. Water Resources Res. 8:533–556.

Young, M. H., P. J. Wierenga, and C. F. Mancino. 1997. Monitoring near-surface soil water storage in turfgrass using time domain reflectometry and weighing lysimetry. Soil Sci. Soc. Amer. J. 61:1138–1147.

Zheng, C. 1990. MT3D: A Modular Three-Dimensional Transport Model for Simulation of Advection, Dispersion and Chemical Reactions of Contaminants in Groundwater Systems. Report to the U.S. Environmental Protection Agency. S.S. Papadopulos & Associates, Inc., Bethesda, MD.

Zheng, C. 1992. PATH3D: A Groundwater Path and Travel-Time Simulator, Version 3.0. S.S. Papadopulos & Associates, Inc., Bethesda, Maryland.

Zwingle, E., and J. Richardson. 1993. Ogallala Aquifer: Wellspring of the High Plains. National Geographic 183: 80–109.

Appendix A

U.S. Geological Survey Programs that Support Groundwater Resources Studies

The U.S. Geological Survey (USGS) has had a central role in characterizing the nation's major aquifers and in developing methods to assess groundwater conditions and processes. Three important concerns about groundwater resources are (1) the sustainability of groundwater resources for long-term water-supply needs, (2) groundwater quality, and (3) environmental effects of groundwater development. Ensuring sustainable groundwater supplies for all needs requires an understanding of subsurface processes in aquifers as well as an understanding of the interaction of groundwater with land and surface water resources. Because groundwater and surface water interact in complex ways, ecosystem studies need to incorporate the effects of this interaction. Thus, knowledge of many fields of study is needed to conduct comprehensive groundwater evaluations. These studies require work at regional and local scales to cover all information needs. Because of its long history of conducting groundwater studies as well as its expertise in hydrology, geology, biology, and mapping, the USGS is in a unique position to provide comprehensive evaluations of groundwater systems.

The USGS has responded to changing groundwater issues by designing and implementing programs that target those issues. For example, the Regional Aquifer-System Analysis (RASA) Program began in response to concerns about groundwater supplies during the 1977 drought. Likewise, the National Water-Quality Assessment (NAWQA) Program began in 1991 in response to concerns about the status and trends in the nation's water quality. These two programs represent USGS responses to a set of groundwater quantity and quality issues.

Water resources needs for ecosystems have also emerged as a new set of concerns that require understanding of both the amount and quality of water needed to sustain life in wetland, stream, and lake environments. Although USGS programs are targeted to fairly specific issues, they inevitably yield data and insights that contribute to the success of programs designed for other issues.

Ground-Water Resources Program

The Ground-Water Resources Program (GWRP) evolved from the RASA Program. From 1978 to 1995, the RASA Program systematically evaluated 25 of the nation's most important groundwater systems. Most of the RASA studies calculated regional groundwater budgets for both pumping and prepumping conditions. During the 1970s and 1980s, pumpage from 11 of the 25 regional aquifer systems provided from 40 percent to 50 percent of the groundwater withdrawn in the United States (Johnston, 1997). Computer models were used to estimate the effects of pumping on water levels. In general, the RASA Program did not examine many of the shallower or less-productive aquifers that are important to ecosystem studies and to many rural and small community water users as well as for sustaining flow in streams. In addition to about 1,100 published reports, the RASA Program published a National Ground-Water Atlas that used RASA data and data from other agencies as a general source of information on groundwater resources (http://capp.water.usgs.gov/gwa/index.html).

The RASA Program, which ended in 1995, evolved into the GWRP, a program that places more emphasis on a broader range of groundwater issues. Current GWRP work consists of four activities:

1. Middle Rio Grande Basin, New Mexico—Studies by the USGS in cooperation with the New Mexico Bureau of Mines and Mineral Resources and the city of Albuquerque have shown that groundwater is not as plentiful as once thought in the Middle Rio Grande basin. Multidivisional efforts to more completely understand this complex hydrogeologic system will culminate in a groundwater-flow model of the area that may be used to help policy makers decide new courses of action for a number of complex groundwater issues.

2. Southwestern United States—Surface water in the southwestern United States is generally fully appropriated, and considerable ground

water is pumped for irrigation and supply. New water supplies increasingly rely on conjunctive use of surface water and groundwater. Sensitive ecosystems also rely on groundwater, a situation that creates further competition for scarce water resources. To address these concerns, the GWRP began a study of the interaction of groundwater and surface water in the Southwest in October 1998.

3. Atlantic Coast—Development of groundwater resources along the Atlantic coast has caused salt water to intrude many highly productive aquifers. Related concerns exist about the effects of changes in groundwater discharge to coastal ecosystems. A project to review what is known about these freshwater–saltwater issues along the Atlantic coast was begun in October 1998.

4. National Aquifer Data Base—Preliminary planning is underway for a digital database on principal aquifer systems as a follow-up to the National Ground-Water Atlas.

The GWRP thus addresses a variety of information needs. As the program transitions from its exclusive focus on the 25 RASA aquifer systems to broader issues, the above activities can serve as prototypes for possible future activities of the Ground-Water Resources Program. It is the main USGS program for assessing issues related to groundwater resources at the regional scale.

National Water-Quality Assessment Program

The National Water-Quality Assessment (NAWQA) Program began in 1991 to provide information about the status and trends in the quality of the nation's groundwater and surface water resources. The focus of the groundwater component of the NAWQA Program has been to determine the effects of human activities on the quality of groundwater in agricultural and urban areas. The program does not evaluate issues such as salt-water intrusion or many of the complex issues associated with the interaction of groundwater with streams, lakes, wetlands, and other surface water bodies. GWRP studies use information from NAWQA studies as well as provide information to them. For example, two of the prototype areas chosen for the National Aquifer Data Base are the High Plains and North Atlantic Coastal Plain aquifer systems. These aquifers are also included in NAWQA study units.

National Cooperative Geologic Mapping Program

The National Cooperative Geologic Mapping Program (NCGMP) was created as a partnership of USGS with the state geological surveys and universities to produce geologic maps and databases. About a third of the projects conducted by this program have cited groundwater issues as one of the principal reasons for mapping. Because there is a geologic component to nearly all groundwater studies, coordination between this program and the GWRP is an important element of groundwater efforts in the USGS. Since 1992, the NCGMP has produced geologic maps and databases as the framework for groundwater assessments of regional aquifers in the Southwest and the southeastern Coastal Plain.

Federal–State Cooperative Water Program

The Federal–State Cooperative Water (Coop) Program matches funds from state and local agencies to support data collection and investigations that serve both federal interests and the needs of the state and local agencies. For over 100 years, this arrangement has been a valuable means of building a national database of information that contains water quality, water-use, surface water, and groundwater data to develop a more complete understanding of hydrologic conditions including groundwater conditions. The RASA and NAWQA Programs have depended greatly on previous work in the Coop Program as the starting point for many of their efforts. Close interaction with water resources officials at the state and local levels ensures continuing relevance of the groundwater studies to address the most pressing issues.

Toxic Substances Hydrology Program

The Toxic Substances Hydrology (Toxics) Program was established in 1982 to provide earth science information on the behavior of toxic substances in the hydrologic environment including groundwater. Most of the research activities take place at contaminated sites, usually at the local scale. A few studies, such as the Midwest pesticide study, are conducted at the regional scale. Hydraulic properties of aquifers and confining units at the local scale are important aspects of groundwater studies for the Toxics Program. These data are also needed for groundwater studies conducted by other USGS programs.

National Research Program

The National Research Program (NRP) conducts fundamental and applied research on hydrologic problems and develops techniques and methodologies for the USGS. Most research conducted by NRP scientists focuses on long-term investigations that integrate hydrologic, geologic, chemical, climatic, and biological information related to water resources and environmental problems. The program is designed to encourage a long-term interdisciplinary approach to solving hydrologic problems. The NRP makes a deliberate effort to anticipate research needs that will be pertinent to the broad hydrologic science issues of the future. Thus, activities of NRP scientists change through time, reflecting the emergence of promising new areas of inquiry and the demand for new tools and techniques to address water resources issues and problems. The direct linkage of the program with other water resources programs of the USGS ensures that the research remains relevant to current water resources needs. NRP scientists interact with all USGS programs in the Water Resources Division (WRD), many programs in other divisions, and researchers at universities nationwide.

Through the programs highlighted above, the USGS has a wide range of capabilities in hydrology, biology, geology, and mapping to address groundwater resources in a fully integrated manner. Other examples include expertise in coastal geology that enhances our understanding of near-shore geologic environments to help with issues such as freshwater discharge at the coasts and their relations to salt-water intrusion into aquifers. Capabilities in remote sensing and land characterization provide key information for computer models and decision-support systems. Expertise in climate improves understanding of the role of climate variability and climate change on groundwater resources. Biological capabilities in habitats, wetlands, and instream-flow requirements are needed to assess the effects of groundwater development on surface water systems. Over the years, USGS groundwater studies have become more interdisciplinary. As water-supply and water quality issues become more intertwined, answers to questions about sustainable supplies of groundwater will be found by considering the interrelations of work done in the fields of hydrology, geology, biology, and geochemistry.

Appendix B

Biographical Sketches of Committee Members

KENNETH R. BRADBURY is a research hydrogeologist/professor with the Wisconsin Geological and Natural History Survey, University of Wisconsin-Extension, in Madison. He received his Ph.D. (hydrogeology, 1982) from the University of Wisconsin-Madison, his A.M. (geology, 1977) from Indiana University, and his B.A. (geology, 1974) from Ohio Wesleyan University. His current research interests include groundwater flow in fractured media, ground water recharge processes, wellhead protection, and the hydrogeology of glacial deposits.

VICTOR R. BAKER is regents professor and head of the Department of Hydrology and Water Resources at the University of Arizona. He is also professor of geosciences and professor of planetary sciences at the University of Arizona. His research interests include geomorphology, flood geomorphology, paleohydrology, Quaternary geology, natural hazards, geology of Mars and Venus, and philosophy of earth and planetary sciences. He has spent time as a geophysicist for U.S. Geological Survey and as an urban geologist. He has served on various committees and panels of the National Research Council, including the Panel on Alluvial Fan Flooding, the Panel on Global Surficial Geofluxes, and the Panel on Scientific Responsibility and Conduct of Research. He formerly chaired the U.S. National Committee for the International Union for Quaternary Research (INQUA) and served on the Global Change Committee Working Group on Solid Earth Processes. Dr. Baker was recently president of the Geological Society of America and president of the INQUA Commission on Global Continental

Paleohydrology. He holds a B.S. from Rensselaer Polytechnic Institute (1967) and a Ph.D. from the University of Colorado (1971).

ANA P. BARROS is an associate professor of civil engineering at Harvard University. Until recently, she was on the faculty at the Pennsylvania State University. She received a diploma in civil engineering from the University of Porto (Portugal) in 1985, an M.S. in hydraulics/ocean engineering from the University of Porto in 1988, an M.S. in environmental science and engineering from Oregon Graduate Institute in 1990, and a Ph.D. in civil engineering from the University of Washington in 1993. Dr. Barros's research interests are environmental fluid mechanics, land-atmosphere interactions, macroscale hydrology, hydrometeorology of mountainous regions, hydrologic extremes (floods and droughts), climate variability, and remote sensing.

MICHAEL E. CAMPANA is director of the Water Resources Program and professor of earth and planetary sciences at the University of New Mexico. His current interests are hydrologic system-aquatic ecosystem interactions, regional hydrogeology, environmental isotope hydrology, and the hydrology of arid and tropical regions. He teaches courses in water resources management, hydrogeology, subsurface fate and transport processes, environmental mechanics, and geological fluid mechanics. He was a Fulbright scholar to Belize in 1996. Dr. Campana received a B.S. in 1970 in geology from the College of William and Mary, an M.S. in hydrology in 1973, and a Ph.D. in hydrology in 1975 from the University of Arizona.

BENEDYKT DZIEGIELEWSKI is an Associate Professor of Geography at Southern Illinois University at Carbondale, and Executive Director of the International Water Resources Association. His two main research areas are Water Demand Management (urban water conservation planning and evaluation, water demand forecasting, modeling of water use in urban sectors) and Urban Drought (drought planning and management, measurement of economic, social and environmental drought impacts). He is Editor-In-Chief of *Water International*, and is an Honorary Lifetime Member of the Water Conservation Committee of the American Water Works Association. He received his B.S. and M.S. in Environmental Engineering from Wroclaw Polytechnic University, Wroclaw, Poland, and his Ph.D. in Geography and Environmental Engineering from Southern Illinois University.

KIMBERLY A. GRAY is an associate professor of environmental engineering in the Department of Civil Engineering at Northwestern University. She received her Ph.D. from Johns Hopkins in 1988, an M.S. from the University of Miami in 1983 in civil engineering, and her B.A. in 1978 in biology from Northwestern University. Dr. Gray teaches physicochemical processes, aquatic chemistry, environmental analytical chemistry, and drinking water treatment design. Her research entails experimental study of both engineered and natural processes. She studies the characteristics of natural organic matter in surface waters, wetlands, and treatment systems by pyrolysis-GC-MS. Other topics of her research include the use of semiconductors to photocatalyze the destruction of hazardous chemicals, the application of ionizing radiation to reductively dechlorinate pollutants in soil matrices, and the ecotoxicology of PCBs in periphytic biolayers.

C. THOMAS HAAN is the regents professor and Sarkeys distinguished professor in the Department of Biosystems and Agricultural Engineering at Oklahoma State University. He received his Ph.D. in agricultural engineering from Iowa State University in 1967. Dr. Haan's research interests are hydrology, hydrologic and water quality modeling, stochastic hydrology, and risk analysis. He has served as a consultant to several national and international agencies. Dr. Haan is a member of the National Academy of Engineering.

DAVID R. MAIDMENT is the Ashley H. Priddy Centennial Professor of Engineering, and Director of the Center for Research in Water Resources at the University of Texas at Austin. He is an acknowledged leader in the application of Geographic Information Systems to hydrologic modeling. His current research involves the application of GIS to flood plain mapping, water quality modeling, water resources assessment, hydrologic simulation, surface water-groundwater interaction, and global hydrology. He is the co-author of *Applied Hydrology* (McGraw-Hill, 1988) and the editor-in-chief of *Handbook of Hydrology* (McGraw-Hill, 1993). From 1992 to 1995 he was Editor of the *Journal of Hydrology*, and he is currently an Associate Editor of that Journal and of the *Journal of Hydrologic Engineering*. He received his Bachelor's degree in Agricultural Engineering from the University of Canterbury, Christchurch, New Zealand, and his M.S. and Ph.D. degrees in Civil Engineering from the University of Illinois at Urbana-Champaign.

Appendix B 141

DAVID H. MOREAU is professor in the Departments of City and Regional Planning and Environmental Sciences and Engineering at the University of North Carolina, Chapel Hill. Chair of the Department of City and Regional Planning, Dr. Moreau received a B.Sc. (civil engineering, 1960) from Mississippi State University, an M.Sc. (civil engineering, 1963) from North Carolina State University, an M.Sc. (engineering, 1964) from Harvard University, and a Ph.D. (water resources, 1967) from Harvard University. Dr. Moreau has been a consultant to the United Nations Development Program, Water Management Models for Water Supply; New York City, review of water demand projections; and Water for Sanitation and Health Program (AID), financing of water supply and waste disposal.

KAREN L. PRESTEGAARD is Associate Professor of Geology at the University of Maryland. Her research interests include sediment transport and depositional processes in mountain gravel-bed streams; mechanisms of streamflow generation and their variations with watershed scale, geology, and land use; hydrologic behavior of frozen ground; hydrologic consequences of climate change; and hydrology of coastal and riparian wetlands. She was a member of the NRC/CGER/-BRWM Committee for Yucca Mountain Peer Review: Surface Characteristics, Preclosure Hydrology, and Erosion. She received her B.A. in geology from the University of Wisconsin-Madison, and her M.S. and Ph.D. in geology from the University of California, Berkeley.

STUART S. SCHWARTZ is a private consultant specializing in water resource systems analysis. He received his B.S. and M.S. in biology-geology from the University of Rochester, and Ph.D. in systems analysis from the Johns Hopkins University. Dr. Schwartz was director of the Section for Cooperative Water Supply Operations on the Potomac (CO-OP) at the Interstate Commission on the Potomac River Basin. His research and professional interests focus on the application of systems analysis and multiobjective optimization in risk-based water resource management.

DONALD I. SIEGEL is a Professor of Geology at Syracuse University where he teaches graduate courses in hydrogeology and aqueous geochemistry. He holds B.S. and M.S. degrees in geology from the University of Rhode Island and Penn State University, respectively, and a Ph.D. in hydrogeology from the University of Minnesota. His research interests are in solute transport at both local and regional scales, wetland-ground water interaction, and paleohydrogeology. He was a member of the NRC's Committee on Techniques for Assessing Ground Water Vulnerability and Committee on Wetlands Characterization.

VERNON L. SNOEYINK is the Ivan Racheff Professor of Environmental Engineering at the University of Illinois. His primary areas of research are the physical and chemical processes for drinking water purification, in particular the removal of organic contaminants by activated carbon adsorption. In 1980, he co-authored the textbook *Water Chemistry* (Wiley and Sons). He has been a trustee of the American Water Works Association Research Foundation and president of the Association of Environmental Engineering Professors. He is now a member of the editorial advisory board of the *Journal of the American Water Works Association* and vice-chair of the Drinking Water Committee of the Environmental Protection Agency's Science Advisory Board. He was elected to the National Academy of Engineering in 1998. He has been a member of several NRC committees, and chaired the Committee on Small Water Supply Systems. He received his B.S. and M.S. degrees in civil engineering and his Ph.D. in water resources engineering from the University of Michigan.

MARY W. STOERTZ is an assistant professor of hydrogeology at Ohio University, Department of Geological Sciences. Her area of specialty is stream restoration, particularly acid mine drainage polluted streams of Appalachia, as well as restoration of channelized rivers. She founded the Appalachian Watershed Research Group at Ohio University, which has the mission of restoring desired functions of watersheds subject to mining, sedimentation and flooding. She directs the miltidisciplinary research arms of the Monday Creek Restoration Project and the Raccoon Creek Improvement Committee. Dr. Stoertz received her B.S. (geology) from the University of Washington, and her M.S. and Ph.D. (hydrogeology, minor in civil and environmental engineering) from the University of Wisconsin-Madison.

KAY D. THOMPSON is assistant professor at Washington University, Department of Civil Engineering. Her research is to investigate properties of subsurface materials for ground water studies, develop methods for subsurface characterization, assess the risks of hydrologic dam failure, and consult on minimizing environmental impacts during development. Dr. Thompson received a B.S. in civil engineering and operations research in 1987 from Princeton University, an M.S. in 1990 from Cornell University, and a Ph.D. in 1994 in civil and environmental engineering from the Massachusetts Institute of Technology.